Basic Math for Process Control

Basic Math for Process Control

by Bob Connell

Notice

The information presented in this publication is for the general education of the reader. Because neither the author nor the publisher have any control over the use of the information by the reader, both the author and the publisher disclaim any and all liability of any kind arising out of such use. The reader is expected to exercise sound professional judgment in using any of the information presented in a particular application.

Additionally, neither the author nor the publisher have investigated or considered the affect of any patents on the ability of the reader to use any of the information in a particular application. The reader is responsible for reviewing any possible patents that may affect any particular use of the information presented.

Any references to commercial products in the work are cited as examples only. Neither the author nor the publisher endorse any referenced commercial product. Any trademarks or tradenames referenced belong to the respective owner of the mark or name. Neither the author nor the publisher make any representation regarding the availability of any referenced commercial product at any time. The manufacturer's instructions on use of any commercial product must be followed at all times, even if in conflict with the information in this publication.

ISA
67 Alexander Drive
P.O. Box 12277
Research Triangle Park, NC 27709

Library of Congress Cataloging-in-Publication Data
  Connell, Bob.
   Basic math for process control / by Bob Connell.
      p. cm.
  Includes bibliographical references and index.
   ISBN 1-55617-813-1
   1. Chemical process control--Mathematics. I. Title.
   TP155.75 .C662 2003
   660'.281--dc21
                              2002013091

Contents

Chapter 10   THE Z–N APPROXIMATION   131

Chapter 11   UNITS, BEST VALUES, FORMULAS, AND OTHER GOOD STUFF   151

INDEX   173

Preface

In order to become proficient in any branch of technology, the knowledge required will have certain building blocks in its foundation. For students aspiring to become knowledgeable in process control, one of the important blocks is mathematics. As such, any student who is striving for the certification necessary to enter the process control field can expect to be subjected to one or more courses in mathematics.

Unfortunately, courses in mathematics tend to be taught by instructors whose mathematical minds are far above those of their students. The same applies to the authors of the textbooks which are dutifully purchased as an adjunct to the class room teaching. What this can mean then, is that courses in mathematics which are intended to lead to a knowledge of process control, can instead become an obstacle to success. The math course has to be passed, after all.

When I was striving to comprehend control theory, I had trouble with the math personally, not so much because it was to deep for me, but because of the way that it was presented. There were just too many gaps in the explanation. Consequently, in this text I have tried to present the mathematical concepts in the way that I wish they had been laid out for me.

In my own mind, I have an admiration and respect for mathematics, because mathematics is basically an exercise in thinking logically. Rules in mathematics are always hard and fast. From my personal observations of the way that many process control situations are dealt with in industry, it is unlikely that there is any branch of technology that is more in need of logical thinking.

I confess, at the outset, that I am in no way an authority on mathematics. My knowledge of mathematics really doesn't go one centimetre beyond what this book covers. In fact, if it were not for a wonderful stroke of luck in which I came into contact with Mrs. Florica Pascal, I would not have been able to complete this text. A superior mathematician in her own right, Mrs. Pascal reviewed chapters, corrected mistakes, and showed me how to solve problems which were beyond my humble capabilities.

All of this means that this text on mathematics was written not by an expert, but by an engineer who has to see, and understand, each step in the development of a mathematical entity. The way the text is written more or less bears this out. An accomplished mathematician will likely find it trivial or boring. But to the student of process control who has to get through the math course which the curriculum requires, it may just prove to be helpful. And, if whatever help was provided carries on into the on-the-job phase, so much the better.

Bob Connell

1

Trigonometry and Cyclic Functions

Trigonometry is a branch of mathematics concerned with functions that describe angles. Although knowledge of trigonometry is valuable in surveying and navigation, in control systems engineering its virtue lies in the fact that trigonometric functions can be used to describe the status of objects that exhibit repeatable behavior. This includes the motion of the planets, pendulums, a mass suspended on a spring, and perhaps most relevant here, the oscillation of process variables under control.

Units of Measurement

The most common unit of measurement for angles is the *degree*, which is 1/360 of a whole circle.

A lesser used unit is the *radian*. Although the radian is not ordinarily used in angular measurement, it should be understood because when differential equations, which occur in control systems engineering, are solved, the angles emerge in radians.

On the circumference of a circle, if an arc equal in length to the radius of the circle is marked off, then the arc will subtend, at the center of the circle, an angle of 1 radian. The angle 0 (or POB) in Figure 1-1, illustrates this.

In line with this definition of a radian, the relationship between radians and degrees can be worked out. The full circumference of the circle (length $2\pi r$) subtends an angle of 360° at the center of the circle. An arc of length r will subtend an angle of

$$\frac{r}{2\pi r} \times 360° = \frac{180}{\pi} \text{ degrees.}$$

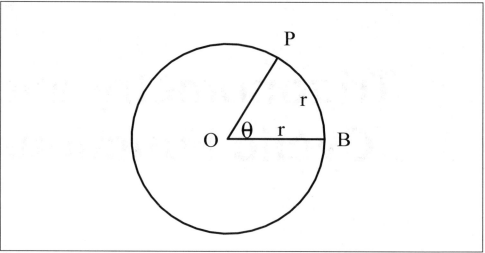

Figure 1-1. A radian defined.

Therefore 1 radian = $\dfrac{180}{\pi}$ deg, or π radians = $180°$.

The actual value of a radian is 57°17′45″, although this value is hardly ever required in control systems analysis.

If the base line OB in Figure 1-1 remains fixed and the radius OP is allowed to rotate counterclockwise around the center O, then the angle θ (or POB) increases. If the starting point for OP is coincident with OB, and OP rotates one complete rotation (or cycle) until it is again coincident with OB, then the angle θ will be 360°. From this it is evident that 1 cycle = 360° = 2π radians.

Functions of Angles

Let θ be any acute angle for which OB is the base, and P be any point on the inclined side of the angle, as in Figure 1-2. A perpendicular from P down to the base OB meets OB at point A.

First, the ratio of any one of the three sides of the triangle POA, to either of the other sides, is a characteristic of the angle θ. In other words, if any of the ratios PA/OP, OA/OP, or PA/OA is known, then the angle θ can be determined from the appropriate tables.

Note that the values of these three ratios do not depend on the position of P. As P moves out along the inclined side of the angle, OP increases, but PA and OA also increase in the same proportion. The values of the ratios depend on the size of the angle θ, but not on the location of P.

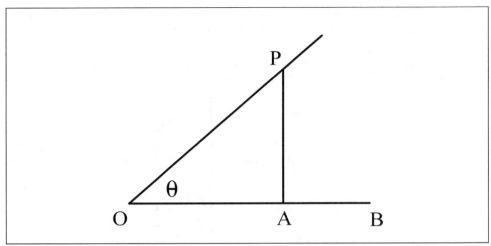

Figure 1-2. Functions of angles.

Definitions

In the triangle POA, the ratio of the length of the side opposite the angle θ to the length of the hypotenuse, or PA/OP, is called the *sine* of the angle θ. The abbreviation *sin* is generally used, that is, PA/OP = sin θ.

The ratio of the side adjacent to the angle θ to the hypotenuse, or OA/OP, is called the *cosine* of the angle θ. This is usually abbreviated *cos*, that is, OA/OP = cos θ.

The ratio of the opposite side to the adjacent side, or PA/OA, is called the *tangent* of the angle θ. This is abbreviated *tan*, so that PA/OA = tan θ.

There are, in addition, some other less common functions of the angle θ. These are defined as follows.

$$\frac{OP}{PA} = \frac{1}{\sin\theta} = \text{cosecant } \theta \text{ (abbreviated cosec } \theta)$$

$$\frac{OP}{OA} = \frac{1}{\cos\theta} = \text{secant } \theta \text{ (abbreviated sec } \theta)$$

$$\frac{OA}{PA} = \frac{1}{\tan\theta} = \text{cotangent } \theta \text{ (abbreviated cot } \theta)$$

Quadrants

The complete circle is divided into four equal parts (called *quadrants*) by horizontal and vertical axes that intersect at the center of the circle. These quadrants are numbered 1 to 4, starting with the upper right quadrant and proceeding counterclockwise, as diagrammed in Figure 1-3.

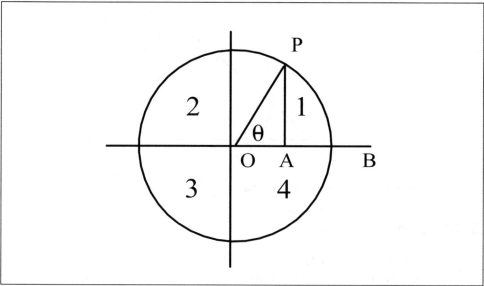

Figure 1-3. The quadrants.

When θ is an acute angle, the radius OP lies in the first quadrant. As the radius rotates counterclockwise and θ increases beyond 90°, OP lies in the second quadrant. For θ between 180° and 270°, OP is in the third quadrant. For θ between 270° and 360°, OP is in the fourth quadrant.

By definition, the measurement of the radius OP is always positive. The measurement OA is defined to be positive when the point A is on the right side of the vertical axis, and negative when A is on the left side of the vertical axis. The measurement PA is defined to be positive when P is above the horizontal axis, and negative when P is below the horizontal axis.

This means that sin θ, cos θ, and tan θ can have positive or negative values depending on the quadrant in which OP lies, which in turn depends on the magnitude of the angle θ. The following values consequently prevail.

When θ = 0°, PA = 0, and OA = OP. Therefore sin θ = 0, cos θ = 1, and tan θ = 0.

When θ = 90°, PA = OP, and OA = 0. Therefore sin θ = 1, cos θ = 0, and tan θ becomes infinite.

When θ = 180°, PA = 0, and OA = OP in magnitude, but OA is negative and consequently OA/OP = –1. Therefore sin θ = 0, cos θ = –1, and tan θ = 0.

When $\theta = 270°$, PA = OP in magnitude, but PA is negative, so PA/OP = –1. Therefore, $\sin \theta = -1$, $\cos \theta = 0$, and $\tan \theta$ becomes infinite in the negative direction. The following table summarizes these points.

Table 1-1. Sine, Cosine and Tangent Functions

Quadrant	θ	$\sin \theta$	$\cos \theta$	$\tan \theta$
1	0° to 90°	0 to 1	1 to 0	0 to ∞
2	90° to 180°	1 to 0	0 to –1	$-\infty$ to 0
3	180° to 270°	0 to –1	–1 to 0	0 to ∞
4	270° to 360° (= 0°)	–1 to 0	0 to 1	$-\infty$ to 0

The values in the table above show that the sine, cosine, and tangent functions repeat themselves with time and are therefore cyclic. *Periodic* is another term that is sometimes used. The sine and cosine functions both cycle between +1 and –1, while the tangent function cycles between $+\infty$ and $-\infty$.

Therefore, sine and cosine functions are useful in describing the behavior of objects and systems that are cyclic. As an example, an object might be known to cycle between the limits of 0 and 10. Its behavior y could be described using the sine function, as

$$y = 5 \sin \theta + 5 = 5 (\sin \theta + 1).$$

In this relationship, when $\sin \theta = 1$, $y = 10$, and when $\sin \theta = -1$, $y = 0$.

Frequency of Cycling

If the motion of an object is linear, the distance traveled is equal to the average velocity of the object multiplied by the elapsed time. In symbol form,

$$s = v\, t$$

where s is in metres, v is in metres per second, and t is in seconds.

The equivalent relation for rotational motion, as in the case of the radius rotating around the center of its circle, is that the angle swept through is equal to its angular velocity multiplied by the elapsed time, or in symbol form,

$$\theta = \omega t$$

where θ is in radians, ω (the Greek letter often used for angular velocity) is in radians per second, and t is in seconds.

Thus, the function sin ωt rather than sin θ may be used to describe the behavior of objects and systems that cycle on a time basis. These include the motion of the planets, pendulums, masses suspended on springs, and the variation with time of temperature, p, and other plant variables that are being controlled.

It is a rule of mathematics that the argument of a sine or cosine function does not have units, or in mathematical terms, it is dimensionless. Consequently if t, which has units of seconds, is in the argument, then the second factor ω must have units that are the inverse of time to balance off the time units. In the basic form, these units will be radians per second. However, radians per second are not particularly appropriate for many cyclic applications. Cycles per second (Hertz, Hz), which are the units for frequency, would be more practical.

If the radius is rotating continuously around its circle, then one complete revolution from zero back to the starting point constitutes one cycle. Its rotational speed is ω radians per second, while its frequency of rotation will be f cycles per second, or Hz.

However, it takes 2π radians to fill out one complete circle, or one cycle. Therefore,

$$f = \frac{\omega}{2\pi} \text{ or } \omega = 2\pi f.$$

Therefore, it is possible to write sin θ = sin ωt = sin 2π f t. In this way, the frequency of the cyclic behavior is identified within the sine function. Functions involving cyclic behavior with time will invariably have a sin ωt or sin (2πf t) term or a cosine term with similar arguments.

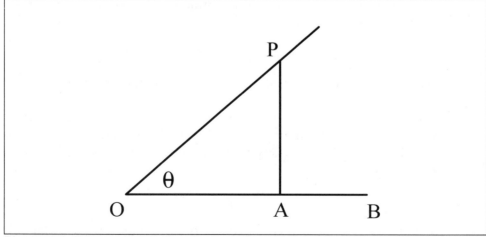

Figure 1-4. Trigonometric function relationships.

Interrelationships

If any one of the trigonometric functions (the sine, cosine, or tangent) is known, then the angle is known. It follows that if any one of the functions is given, the others can be found from it. Consequently, relationships must exist between the various trigonometric functions.

In Figure 1-4,

1. $\tan\theta \;=\; \dfrac{PA}{OA} \;=\; \dfrac{PA}{\dfrac{OP}{OA}} \;=\; \dfrac{\sin\theta}{\cos\theta}$
$\phantom{1. \qquad \tan\theta = \dfrac{PA}{OA} = \dfrac{PA}{}}\overline{OP}$

2.   Since OAP is a right angle triangle, it follows that,

$$(OP)^2 \;=\; (PA)^2 + (OA)^2.$$

Therefore,

$$\frac{(PA)^2}{(OP)^2} + \frac{(OA)^2}{(OP)^2} \;=\; 1 \ \text{ and } \left(\frac{PA}{OP}\right)^2 + \left(\frac{OA}{OP}\right)^2 \;=\; 1$$

Thus, $\sin^2\theta + \cos^2\theta \;=\; 1$.

One relationship that is *not* valid is that the sine of the sum of two angles is equal to the sum of the sines of the individual angles. In other words, if the angles are called X and Y, then sin (X + Y) is *not* equal to sin X + sin Y. The same is true for the cosine and tangent functions.

Sine of the Sum of Two Angles

In Figure 1-5, the angle POA is the sum of an angle X and another angle Y. PC is a perpendicular from point P to the common side OC of the angles X and Y. CD is a perpendicular from point C to the side PA at D. CE is the perpendicular from point C to the extension of OA at point E.

From Figure 1-5, $\sin (X + Y) \;=\; \sin (\text{angle POA}) \;=\; \dfrac{PA}{OP}$.

Developing this further, in triangles OFA and PFC,

Angle OFA = angle PFC,

Angle OAF = angle FCP = 90°.

Therefore, angle FPC = angle FOA = angle X.

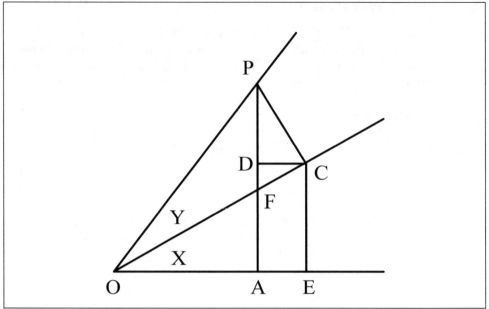

Figure 1-5. Sum of two angles.

$$\frac{PA}{OP} = \frac{PD+DA}{OP} = \frac{CE+PD}{OP} = \frac{CE}{OP}+\frac{PD}{OP} = \left[\frac{CE}{OP}\times\frac{OC}{OC}\right]+\left[\frac{PD}{OP}\times\frac{PC}{PC}\right]$$

$$= \left[\frac{CE}{OC}\times\frac{OC}{OP}\right]+\left[\frac{PD}{PC}\times\frac{PC}{OP}\right]$$

Consequently, $\dfrac{PD}{PC} = \cos X$. In addition,

$$\frac{PC}{OP} = \sin Y, \frac{CE}{OC} = \sin X, \text{ and } \frac{OC}{OP} = \cos Y.$$

Therefore,

$$\sin(X+Y) = \frac{PA}{OP} = \left[\frac{CE}{OC}\times\frac{OC}{OP}\right]+\left[\frac{PD}{PC}\times\frac{PC}{OP}\right] = \sin X\cos Y + \cos X\sin Y.$$

Cosine of a Sum

In Figure 1-5, $\cos(X+Y) = \cos(\text{angle } POA) = \dfrac{OA}{OP}$.

$$\frac{OA}{OP} = \frac{OE-AE}{OP} = \frac{OE}{OP}-\frac{DC}{OP} = \left[\frac{OE}{OP}\times\frac{OC}{OC}\right]-\left[\frac{DC}{OP}\times\frac{PC}{PC}\right]$$

$$= \left[\frac{OE}{OC}\times\frac{OC}{OP}\right]-\left[\frac{DC}{PC}\times\frac{PC}{OP}\right] = \cos X\cos Y - \sin X\sin Y$$

Sine of a Difference Between Two Angles

The sine of the difference between two angles can be determined in a similar manner.

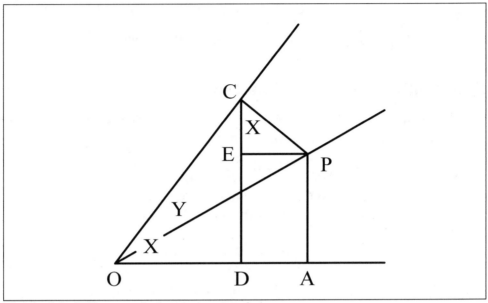

Figure 1-6. Difference between two angles.

In Figure 1-6, the angle POA is the difference between an angle X (COA) and another angle Y (COP). PC is the perpendicular from P to the common side OC of angles X and Y. CD is the perpendicular from C to OA. PE is the perpendicular from P to CD.

Examining this further,

$$\frac{PA}{PO} = \frac{ED}{PO} = \frac{CD - CE}{PO} = \frac{CD}{PO} - \frac{CE}{PO}$$

$$\left[\frac{CD}{PO} \times \frac{OC}{OC}\right] - \left[\frac{CE}{PO} \times \frac{CP}{CP}\right] = \left[\frac{CD}{OC} \times \frac{OC}{PO}\right] - \left[\frac{CE}{CP} \times \frac{CP}{PO}\right]$$

which is equal to $\sin X \cos Y - \left[\frac{CE}{CP} \times \sin Y\right]$.

Triangle ODC is a right angle triangle, therefore Angle COD + angle OCD = 90° = Angle OCD + angle ECP, from which Angle ECP = angle COD = X.

$\frac{CE}{CP}$ = cos (angle ECP) = cos X, and sin (X – Y) = sin X cos Y - cos X sin Y.

Cosine of a Difference

In Figure 1-6, $\cos(X - Y) = \cos(\text{angle POA}) = \dfrac{OA}{OP}$.

$$\frac{OA}{OP} = \frac{OD + DA}{OP} = \frac{OD + EP}{OP} = \frac{OD}{OP} + \frac{EP}{OP} = \left[\frac{OD}{OP} \times \frac{OC}{OC}\right] + \left[\frac{EP}{OP} \times \frac{CP}{CP}\right]$$

$$= \left[\frac{OD}{OC} \times \frac{OC}{OP}\right] + \left[\frac{EP}{CP} \times \frac{CP}{OP}\right] = \cos X \cos Y + \sin X \sin Y$$

Potentially Useful Relationships

The following three relationships often figure in the resolution of problems and, as such, are worth remembering.

1. $\sin^2 X + \cos^2 X = 1$ (developed previously).

2. $\sin 2X = \sin(X + X) = \sin X \cos X + \cos X \sin X = 2 \sin X \cos X.$
       A useful variation of this is

$$\sin X = 2 \sin \frac{X}{2} \cos \frac{X}{2}.$$

3. $\cos 2X = \cos(X + X) = \cos X \cos X - \sin X \sin X = \cos^2 X - \sin^2 X.$

Example 1: Tan (X + Y)

Evaluate $\tan(X + Y)$.

$$\tan(X + Y) = \frac{\sin(X + Y)}{\cos(X + Y)} = \frac{\sin X \cos Y + \cos X \sin Y}{\cos X \cos Y - \sin X \sin Y}$$

$$= \frac{\dfrac{\sin X \cos Y}{\cos X \cos Y} + \dfrac{\cos X \sin Y}{\cos X \cos Y}}{\dfrac{\cos X \cos Y}{\cos X \cos Y} - \dfrac{\sin X \sin Y}{\cos X \cos Y}} = \frac{\tan X + \tan Y}{1 - \tan X \tan Y}$$

Example 2: sin 2X – sin 2Y

Show that $(\sin 2X - \sin 2Y) = 2 \cos(X + Y) \sin(X - Y)$.

Evaluating the right side of the equation:

$$RS = 2[(\cos X \cos Y - \sin X \sin Y)(\sin X \cos Y - \cos X \sin Y)]$$

$$= 2\left[\begin{array}{l}(\cos X \cos Y)(\sin X \cos Y) - (\cos X \cos Y)(\cos X \sin Y) \\ -(\sin X \sin Y)(\sin X \cos Y) + (\sin X \sin Y)(\cos X \sin Y)\end{array}\right]$$

$$= 2 \left[\begin{array}{l} (\sin X \cos X)\cos^2 Y - (\sin Y \cos Y)\cos^2 X - (\sin Y \cos Y)\sin^2 X \\ + (\sin X \cos X)\sin^2 Y \end{array} \right]$$

$$= 2[(\sin X \cos X)(\cos^2 Y + \sin^2 Y) - (\sin Y \cos Y)(\cos^2 X + \sin^2 X)]$$

$$= 2\sin X \cos X - 2\sin Y \cos Y = \sin 2X - \sin 2Y.$$

Example 3: Radius of the Earth

Many people believe that the fact that the earth is round was first pro-
moted by Christopher Columbus, and that on the basis of this knowledge
he sailed off westward to discover the new world. In reality, the credit
should go to a Greek scholar, Erathosthenes, who lived in the third cen-
tury B.C.

As the story goes, about 800 km from Alexandria, Egypt, a shaft had been
dug that was exactly plumb. This shaft was used by astronomers as a
check on the calendar, since there was just one day out of the entire year in
which one could stand at the bottom of the shaft and see the sun. Erathosthenes knew about the shaft.

On the specific day that the sun's rays lit the bottom of the shaft, Erathosthenes was in Alexandria and he noticed that a statue was casting a
shadow. This was somewhat surprising because if the surface of the earth
were flat, there should be no shadow at all. The only plausible explanation
was that the earth's surface, or at least that portion of it between Alexandria and the shaft, must be curved.

Erathosthenes measured the height of the statue and the length of the
shadow. Since these two measurements formed the tangent of the angle
between the statue and the rays of the sun that fell upon it, he could then
calculate the angle, which was 7°.

From Figure 1-7 (out of proportion), it can be seen that the angle formed
by the statue and the sun's rays is the same as the angle subtended by the
two radii emanating from the center of the arc to the shaft and the statue.
This would be dependent on the correctness of the assumption that the
sun is so much larger than the earth that its rays are all parallel when they
arrive at the earth, which is reasonable enough.

It was then only a matter of applying the relation: a (arc) = r × θ (radians).
The arc is 800 km. θ is 7° or 0.122 radians. From this, r = 6550 km.

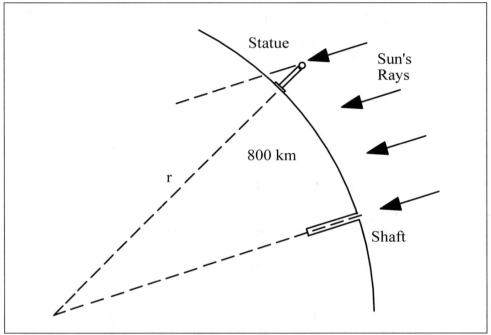

Figure 1-7. Erathosthenes' brilliant deduction.

Using more sophisticated modern techniques, the mean radius of the earth has been computed to be 6371 km, so the error in Erathosthenes calculation was less than 3%. As an afterthought to the story, Columbus likely knew about this gem of knowledge, which was brought to light 17 centuries previously by Erathosthenes. In fact, in the time of Columbus, it is probable that many educated people knew that the earth had to be round, not flat.

Example 4: The Third Side

In the triangle in Figure 1-8, the lengths of two of the sides, a and b, are known, as is the angle θ between them. What is required is to develop an expression involving a, b, and θ, from which the length of the third side, c, can be calculated.

In the triangle, draw a perpendicular from the apex down to the base. This divides the base a into two segments, z_1 and z_2, so that $z_1 + z_2 = a$. The length of the perpendicular is z_3.

$$\frac{z_1}{b} = \cos \theta, \text{ so that } z_1 = b\cos \theta. \text{ Similarly, } z_3 = b\sin \theta$$

$$z_2 = a - z_1 = a - b\cos\theta$$

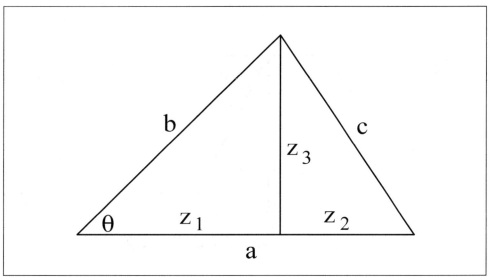

Figure 1-8. Third side problem.

$$c = \sqrt{z_2^2 + z_3^2} = \sqrt{(a - b\cos\theta)^2 + (b\sin\theta)^2}$$

$$= \sqrt{a^2 - 2ab\cos\theta + b^2\cos^2\theta + b^2\sin^2\theta}$$

$$= \sqrt{a^2 - 2ab\cos\theta + b^2(\sin^2\theta + \cos^2\theta)}$$

$$= \sqrt{a^2 + b^2 - 2ab\cos\theta}$$

Example 5: Gain and Phase Lag

This example is potentially useful in designing a computer program that will determine the gain and phase lag in a closed loop control system. In Figure 1-9, OA is a vector whose magnitude is G units and whose orientation with respect to the horizontal axis is the angle a. AB is a vector that is 1 unit in length and has no phase angle. It always lies parallel to the horizontal axis and points in the positive direction.

OB is a vector that is the vector sum of the vector G and the unit vector. The objective is to develop two expressions that a computer can use to calculate the magnitude G_1 and phase angle a_1 of the vector OB, if given the magnitude of G and its phase angle a.

When writing a computer program of this type, it should be kept in mind that computer programs that do mathematics deal with angles in units of

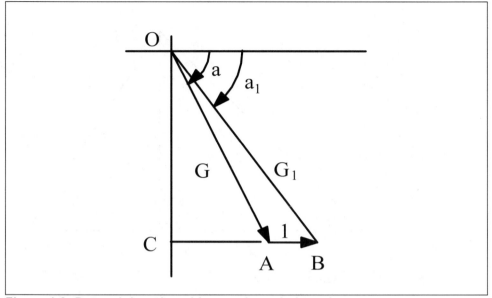

Figure 1-9. Determining closed loop gain and phase lag.

radians, not degrees. Thus, if angle a is specified in degrees, it must be multiplied by the factor 0.0174 5 (π/180).

In the diagram, the angle OAC is equal to angle a.

$$\frac{OC}{G} = \sin a$$

Therefore, $OC = G \sin a$. Similarly, $CA = G \cos a$.

$$CB = CA + 1$$

$$G_1 \text{ magnitude} = \sqrt{(OC)^2 + (CB)^2} = \sqrt{(G \sin a)^2 + (G \cos a + 1)^2}$$

$$a_1 = \text{Angle whose tangent is } \frac{OC}{CB} = \tan^{-1}\left(\frac{G \sin a}{G \cos a + 1}\right)$$

Once again, when the computer calculates the value of the angle a_1, the units of a_1 will be radians not degrees.

2

Differential Calculus

Mathematical relationships are constructed around variable quantities (called variables for short). The relationship shows the way that the value of one of the variables changes when the values of the other variables change.

This implies that in each relationship there is one variable whose value is dependent on the values of the other variables. It is consequently called the *dependent variable*, while the other variables are called the *independent variables*.

The way in which some mathematical relationships are structured often leaves doubt as to which of the variables is the dependent variable. The question becomes more difficult to answer as the number of variables in the relationship increases.

In the control systems engineering field, however, many relationships contain only two variables. Furthermore, one of the variables will be time (designated t). Since a unique characteristic of time is that it pursues its uniform and relentless course into eternity, unaffected by anything else, it is obvious that time cannot be dependent on any other physical variable. Consequently, in all control systems relationships that involve time, it must be the independent variable.

The relationships that are most common in control systems engineering generally show how some dependent variable, which could be distance, temperature, pressure, and so on, varies with time. If the dependent variable is represented by x, then it can be stated that "x is some function of time." In the shorthand of mathematics, this is written $x = f(t)$.

Notice that what this shorthand relationship is telling us is not only that the value of x is dependent on the value of t, but perhaps more important, that the value of x depends *only* on the value of t, and not on the value of any other variable, such as n, y, θ, etc., whatever these characters may represent in the real world. If a number of functions were involved, these might be distinguished by labeling them $f_1(t)$, $f_2(t)$, and so on.

It will often be useful to plot the values of the dependent variable over a range of values of time. When making the plot, it is customary to plot the values of the independent variable (t) along the horizontal axis and the corresponding values of the dependent variable (x, θ, or whatever) along the vertical axis. Figure 2-1 is an example. If the resulting plot is a curve that has no breaks or gaps, then the function is said to be *continuous*. The relationships that arise in control systems engineering can generally be counted on to be *single valued*, meaning that for each value of t there is only one value of the dependent variable.

Concept of Approaching a Limit

When we are first introduced to mathematics, we get the impression that it is an exact science. The rules of mathematics have no exceptions. Everything is based on the concept that something must be equal to something else.

Later, our confidence is shaken when we learn that mathematicians are also concerned with not only what a particular variable is equal to, but also what value that variable may be approaching. This may occur when some other variable, on which the first variable is dependent, approaches its limiting value, which often proves to be zero. Sometimes this apparently nebulous procedure is justified to get around certain complications, one of which might be division by zero. For example, suppose that

$$x = f(t) = \frac{t^2 - t - 6}{t - 3}.$$

What will be value of x when t approaches 3? The function is undefined for t = 3 since division by zero does not compute. However, it turns out that if the denominator (t – 3) is divided into the numerator, the result is (t + 2). This being the case, the value of x *approaches* 5, not infinity, as t *approaches* 3.

A small change (increment) in a variable such as t is usually called *delta t* and is written Δt. While Δt is a small change, it is nevertheless measurable, in whatever units are appropriate. As Δt is made smaller and smaller, a point is reached where we have to ask, "How much smaller can it get without being zero?" In the mathematical sense, it is at this point where Δt becomes the *differential of t*, which is identified by dt. In mathematics shorthand, the limit of the ratio Δx/Δt as Δt approaches zero is written

$$\lim_{\Delta t \to 0} \frac{\Delta x}{\Delta t}.$$

This limit, if it exists, becomes dx/dt, and it is termed *the derivative of x with respect to t*. The function x = f(t) is then said to be differentiable. The ability to differentiate the function f(t) will require not only that the function be continuous, but also that when it is plotted, there are no corners in the graph.

When dealing with increments and with differentials in the determination of limiting values of variables and functions, there is a basic rule with which one need to be familiar. This rule says that in the determination of the limiting value of a derivative, as the value of the increment of the independent variable approaches zero, products and powers of incrementals become insignificant and can be discounted. Specifically, if terms such as Δx^2, Δt^2, or ($\Delta x \Delta t$) emerge, then they can be ignored. This rule must be on board as other derivative functions are developed.

Figure 2-1 is a graph that shows the variation of a dependent variable x with values of the independent variable t. For any specified value of t within the range of the graph, there will be a corresponding value of x, as shown by point P in the graph. Suppose that the value of t increases a small amount. Once again in mathematical shorthand, this small change in t is generally designated Δt. The increase in t will cause a change in x, which will be designated Δx. Depending on the nature of the function f (t), Δx may be positive or negative. Point Q is the new point whose coordinates are t + Δt, and x + Δx.

From the graph it can be seen that the ratio Δx/Δt is the slope of a line that passes through the points P and Q. Furthermore, if Δt, and consequently Δx, were made smaller and smaller, then point Q will approach point P, and the slope of the line through points P and Q will approach the slope of the graph x = f (t) at point P.

The slope of the graph x = f (t) at point P can actually be obtained by determining the value of the ratio Δx/Δt as Δt approaches zero. When Δt is made smaller and smaller, point Q will approach point P, and the slope of the line through points P and Q will approach the slope of the tangent line to the graph x = f(t) at point P. Ultimately, the slope of the graph x = f(t) at point P will be equal to the rate of change of the dependent variable x with t at point P.

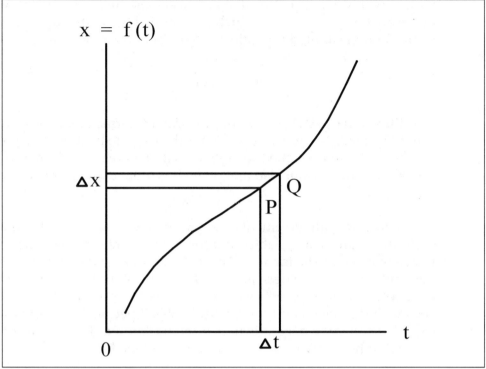

Figure 2-1. A Differentiable Function.

Example 1: f(t) = t²

Suppose that $x = f(t) = t^2$.

Starting from any point t, x, as t changes by an amount Δt, x will change by an amount Δx, which produces a new point $t + \Delta t$, $x + \Delta x$. However, this new point is also on the graph $x = t^2$.

Therefore, $(x + \Delta x) = (t + \Delta t)^2$.

Expanding this, $x + \Delta x = t^2 + 2t\,\Delta t + (\Delta t)^2$.

Since $x = t^2$, these two items can be subtracted from the left and right sides, respectively, of the equation.

Therefore, $\Delta x = 2t\,\Delta t + (\Delta t)^2$, and dividing through by Δt,

$$\frac{\Delta x}{\Delta t} = 2t + \Delta t.$$

In the example above, as Δt approaches zero, the equation $\dfrac{\Delta x}{\Delta t} = 2t + \Delta t$ consequently becomes $\dfrac{dx}{dt} = 2t$.

An important characteristic of the derivative expression is that it can identify the slope of the graph $x = f(t)$ at any point established by a selected value of t. Therefore, the slope of the curve $x = t^2$ will be equal to 2 t anywhere along the curve.

Procedure for Determining a Derivative

The general procedure for determining a derivative expression is to substitute $(x + \Delta x)$ and $(t + \Delta t)$ in the relation $x = f(t)$, evaluate the ratio $\Delta x / \Delta t$, and then determine the limiting value of $\Delta x / \Delta t$ as Δt approaches zero. The limiting value thus determined will be dx/dt, the derivative of x with respect to t.

Derivative of a Sum or Difference

Suppose that x is the sum of two differentiable functions designated u and v, in which $u = f_1(t)$, and $v = f_2(t)$. Since x, u, and v are all functions of t, when t becomes $(t + \Delta t)$, x becomes $x + \Delta x$, u becomes $u + \Delta u$, and v becomes $v + \Delta$, so that $(x + \Delta x) = (u + \Delta u) + (v + \Delta v)$. Since $x = u + v$, they will drop out of this equation leaving $\Delta x = \Delta u + \Delta v$.

Thus, $\Delta x = \Delta u + \Delta v$. Dividing by Δt, $\dfrac{\Delta x}{\Delta t} = \dfrac{\Delta u}{\Delta t} + \dfrac{\Delta v}{\Delta t}$.

As Δt approaches zero, the $\lim\limits_{\Delta t \to 0} \dfrac{dx}{dt} = \dfrac{du}{dt} + \dfrac{dv}{dt}$.

What this reveals is really a basic rule, namely, that the derivative of the sum of two functions is equal to the sum of the individual derivatives. This rule can be extended, without difficulty, to show that the derivative with respect to t (or any other independent variable) of the sum or difference of any number of functions of t, is equal to the sum or difference of the individual derivatives.

Derivative of a Product

Suppose that $x = u \times v$, where x, u, and v are all differentiable functions of the independent variable t. With a small change in t, t becomes $t + \Delta t$. This results in incremental changes in x, u, and v, so that $x + \Delta x = (u + \Delta u)(v + \Delta v) = uv + u\,\Delta v + v\,\Delta u + \Delta u\,\Delta v$.

Since $x = uv$, $\Delta x = u\,\Delta v + \Delta u (v + \Delta v)$.

Dividing by Δt, $\dfrac{\Delta x}{\Delta t} = u \dfrac{\Delta v}{\Delta t} + \dfrac{\Delta u}{\Delta t}(v + \Delta v)$.

The function $v = f_1(t)$ is differentiable, so that as Δt approaches zero, Δv approaches zero. Therefore,

$$\lim_{\Delta t \to 0} \frac{\Delta x}{\Delta t} = \lim_{\Delta t \to 0} \frac{\Delta uv}{\Delta t} = \frac{dx}{dt} = u\frac{dv}{dt} + v\frac{du}{dt}.$$

Derivative of a Quotient

As before, x, u, and v are all differentiable functions of the independent variable t, and it is given that $x = u/v$. If t should change incrementally to $t+\Delta t$, x will become $x + \Delta x$, which will be equal to

$$\frac{u + \Delta u}{v + \Delta v}.$$

Therefore,

$$\Delta x = \frac{u + \Delta u}{v + \Delta v} - x = \frac{u + \Delta u}{v + \Delta v} - \frac{u}{v} = \frac{uv + v\Delta u - uv - u\Delta v}{v(v + \Delta v)} = \frac{v\Delta u - u\Delta v}{v(v + \Delta v)}.$$

Dividing both sides by Δt:

$$\frac{\Delta x}{\Delta t} = \frac{v\dfrac{\Delta u}{\Delta t} - u\dfrac{\Delta v}{\Delta t}}{v(v + \Delta v)}.$$

As Δt approaches zero, Δv approaches zero. Consequently,

$$\lim_{\Delta t \to 0} \frac{\Delta x}{\Delta t} = \frac{dx}{dt} = \frac{v\dfrac{du}{dt} - u\dfrac{dv}{dt}}{v^2}.$$

Dimensions and Units

A requirement of mathematics is that the argument of a trigonometric function, or the exponent of an exponential function, be dimensionless, that is, it should not have units. Specifically, in functions such as sin θ, cos θ, and e^{θ}, the variable θ must not have units associated with it.

This is why it is not appropriate to use the variable t as the argument or exponent in trigonometric or exponential functions because in control systems studies, t usually stands for time, and time has definite units of seconds, minutes, or hours. If time does appear in the argument or exponent of functions of these types, then it has to be compensated for by coupling it with a second variable such as omega (ω). For the combination ωt to be dimensionless, the units of ω must be the inverse of time. If the units of ω are radians per second or cycles per second, this will meet the requirement because angles in radians and cycles or revolutions have no units, and the "per second" dimension in the ω compensates for the "seconds" in the t.

Derivative of a Sine Function

Given the function f (θ) = x = sin θ, as θ becomes (θ + Δθ), sin (θ + Δθ) causes x to become (x + Δx). Accordingly,

$$x + \Delta x = \sin (\theta + \Delta\theta) = \sin \theta \cos \Delta\theta + \cos \theta \sin \Delta\theta$$

$$\Delta x = \sin \theta \cos \Delta\theta + \cos \theta \sin \Delta\theta - x = \sin \theta \cos \Delta\theta + \cos \theta \sin \Delta\theta - \sin \theta.$$

It is a fact of mathematics that when the magnitude of an angle (θ) approaches zero, then the value of sin θ approaches the value of θ. The numbers in the table following bear this out.

Table 2-1. Magnitude of an Angle (θ)

θ degrees	θ radians	sin θ	cos θ
10	0.1745	0.1736	0.9848
8	0.1396	0.1392	0.9903
6	0.1047	0.1045	0.9945
4	0.0698	0.0698	0.9976
2	0.0349	0.0349	0.9994

Also, as θ approaches zero, cos θ approaches 1. Therefore, as Δθ approaches zero, Δx approaches sin θ + cos θ (Δθ) – sin θ = cos θ(Δθ), and Δx/Δθ approaches cos θ.

Thus the derivative $\dfrac{d}{d\theta} \sin \theta = \cos \theta$.

Binomial Theorem

As a prerequisite to obtaining the expression for the derivative of the independent variable raised to a power other than one, some familiarity with the binomial theorem and the expansion for the power of a sum is helpful. Some of the terms in this expansion contain factorial quantities. A factorial applies only to positive integers, and in mathematical shorthand, the factorial operator consists of an exclamation mark (!) following the integer.

The factorial of a number (positive integer) is equal to the number multiplied by all of the numbers less than itself, in sequence, down to the number one. Thus, factorial 5 would be 5! = 5 × 4 × 3 × 2 × 1 = 120.

In general, n! = n × (n – 1) × (n – 2) × ... × 3 × 2 × 1, assuming that in this statement, (n – 2) is a number larger than 3.

The expansion for the expression $(a + b)^n$ using the binomial theorem is:

$$(a + b)^n = a^n + na^{n-1}b + \frac{n(n-1)}{2!}a^{n-2}b^2 + \frac{n(n-1)(n-2)}{3!}a^{n-3}b^3 + \ldots$$

The expansion ends with the term that has $(a^{(n-n)} = a^0 = 1) \times b^n$.

As a test of the theorem, set n = 4. Then,

$$(a + b)^4 = a^4 + 4a^3b + \frac{4 \times 3}{2 \times 1}a^2b^2 + \frac{4 \times 3 \times 2}{3 \times 2 \times 1}ab^3 + \frac{4 \times 3 \times 2 \times 1}{4 \times 3 \times 2 \times 1}b^4$$

$$= a^4 + 4a^3b + 6a^2b^2 + 4ab^3 + b^4$$

which is the same result as would be obtained by multiplying (a + b) by (a + b) three times algebraically.

Derivative of a Power

Given that $x = t^n$, if t changes to t + Δt, then x becomes x + Δx, and the new relation is $(x + \Delta x) = (t + \Delta t)^n$. Using the binomial theorem to expand $(t + \Delta t)^n$,

$$x + \Delta x = (t + \Delta t)^n$$

$$= t^n + nt^{n-1}\Delta t + \frac{n(n-1)}{2!}t^{n-2}\Delta t^2 + \frac{n(n-1)(n-2)}{3!}t^{n-3}\Delta t^3 + \ldots + \Delta t^n.$$

Since $x = t^n$, x can be removed from the left side, and t^n from the right.

Dividing by Δt,

$$\frac{\Delta x}{\Delta t} = nt^{n-1} + \frac{n(n-1)}{2!}t^{n-2}\Delta t + \frac{n(n-1)(n-2)}{3!}t^{n-3}\Delta t^2 + \ldots + \Delta t^{n-1}.$$

As Δt approaches zero, all of the terms on the right side except the first term approach zero because they are multiplied by some positive power of Δt. Therefore the limit condition is

$$\frac{d}{dt}t^n = \lim_{\Delta t \to 0}\frac{\Delta x}{\Delta t} = \frac{dx}{dt} = nt^{n-1}.$$

Thus, if $x = t^3$, then $\dfrac{dx}{dt} = 3t^2$.

This relationship is also valid if n is a fraction. Suppose that $x = \sqrt{t} = t^{\frac{1}{2}}$. Then,

$$\frac{dx}{dt} = \frac{1}{2}t^{-\frac{1}{2}} = \frac{1}{2}\frac{1}{\sqrt{t}}.$$

In fact, n can be any real number.

Derivative of an Exponential

Given that $x = e^p$, what will be the value of $\dfrac{dx}{dp}$?

The easiest way to determine the derivative is to express e^p in its power series form. (This is developed in Chapter 4.)

$$e^p = 1 + p + \frac{p^2}{2!} + \frac{p^3}{3!} + \frac{p^4}{4!} + \dots$$

Taking the derivative of the series, term by term,

$$\frac{dx}{dp} = 0 + 1 + \frac{2p}{2 \times 1} + \frac{3p^2}{3 \times 2 \times 1} + \frac{4p^3}{4 \times 3 \times 2 \times 1} + \dots = 1 + p + \frac{p^2}{2!} + \frac{p^3}{3!} + \dots$$

which is the original function. Consequently,

$$\frac{d}{dp}e^p = e^p.$$

The power series for sin θ and for cos θ are both developed in Chapter 4. By taking the derivative of the series for sin θ, term by term, it becomes the series for cos θ, which verifies that the derivative of sin θ is cos θ. If the derivative of the series for cos θ is taken, term by term, it becomes the series for sin θ, multiplied by (–1). Hence the derivative of cos θ is –sin θ.

A table of derivatives of selected functions of t is on page 29.

A Function Within a Function

In working out derivative expressions, there is a trap into which an unwary student can fall, and it is associated with taking the derivative of a function that has a second function within it.

If $x = \sin\theta$, then $\dfrac{dx}{d\theta} = \cos\theta$. But if $x = \sin\theta^2$, $\dfrac{dx}{d\theta}$ is not equal to $\cos\theta^2$.

In the second case, θ^2 is not θ but a function of θ within the sine function.

Similarly, if $x = e^p$, then $\dfrac{dx}{dp} = e^p$; but if $x = e^{ap}$, $\dfrac{dx}{dp}$ is not equal to e^{ap}.

Even though $a \times p$ is the simplest possible function of p, it is nevertheless a function.

The first step in dealing with the problem of the derivative of a function that has another function imbedded in it is to recognize that there is a complication, avoiding the trap described. Then, the procedure is to introduce an intermediate variable (u or whatever) and set it equal to the function inside of the base function. The required derivative can then be worked out by using the relationship

$$\frac{dx}{dp} = \frac{dx}{du} \times \frac{du}{dp}.$$

Given that $x = \sin\theta^2$, what is the derivative?

Designate the intermediate function $u = \theta^2$. Then,

$$\frac{du}{d\theta} = 2\theta.$$

Also, $x = \sin u$, and $\dfrac{dx}{du} = \cos u$.

$$\frac{dx}{d\theta} = \frac{dx}{du} \times \frac{du}{d\theta} = \cos u \times 2\theta = 2\theta\cos\theta^2.$$

Similarly, if $x = e^{at}$, designate $u = at$.

$$\frac{du}{dt} = a. \text{ Also } x = e^u, \text{ and } \frac{dx}{du} = e^u.$$

$$\frac{dx}{dt} = \frac{dx}{du} \times \frac{du}{dt} = e^u \times a = ae^{at}.$$

Example 2: Detecting a Maximum or Minimum

One use of the derivative function is that it can often detect the presence of a maximum or minimum point within the original function. This is due to the fact that when the function passes through a maximum or minimum, the slope of the tangent to the curve is horizontal, and the value of the derivative is zero. Consider the function

$$y = \frac{6x - x^2 - 1}{4}.$$

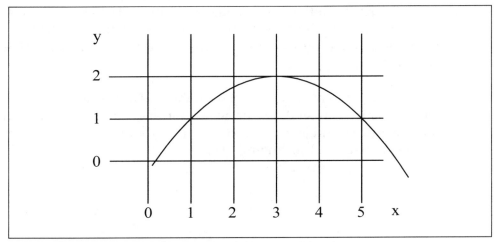

Figure 2-2. Graph of the function $y = \frac{1}{4}(6x - x^2 - 1)$.

This is a parabolic function in the variable x. As can be seen from Figure 2-2, it peaks at x = 3. The peak value is y = 2. The peak point and its value can be calculated by taking the derivative with respect to x and equating it to zero.

$$\frac{dy}{dx} = \frac{1}{4}(6 - 2x)$$

from which (6 – 2x) = 0, and x = 3.

Substituting x = 3 into the original function,

$$y = \frac{18 - 9 - 1}{4} = 2.$$

Example 3: Watering the Lawn

In another example, suppose that you wish to water your lawn, but your garden hose will stretch only to the midpoint of the lawn. You know from experience that to get the water to reach farther, you need to tip the nozzle upward. Intuition tells you that the water will reach the farthest point possible in the horizontal direction if you tilt the nozzle upward at an angle of 45°. At an angle less than 45°, the water jet falls short. If the angle is greater than 45°, the jet travels upward rather than outward and again falls short. The optimum angle appears to be 45°, but can this be proven?

In Figure 2-3, the nozzle is tilted upward at an angle θ to the horizontal. The water jets from the nozzle with a velocity w. The horizontal component of the velocity is w cos θ. Thus the distance S the jet will travel outward will be the product of its velocity and its time of travel, that is S = wcosθ × t.

The value of t can be determined from the upward component of the velocity, which is w sinθ. The basic relation is v = u + at, where u is the initial velocity of the object, v is its final velocity, and a is its acceleration over the timed interval. Rearranging this,

$$t = \frac{v - u}{a}.$$

Figure 2-3. The optimum angle (© Washington Post Writers Group).

For the jet of water, the initial velocity is w sinθ, assuming that velocities in the upward direction are positive. The acceleration due to gravity (–g) will cause the velocity of the jet to decline to zero and then fall back to earth. Its velocity on returning will be the same as when it started out, but in the opposite direction, that is, $-w \sin\theta$. Accordingly,

$$t = \frac{v - u}{a} \text{ becomes } \frac{(-w\sin\theta) - w\sin\theta}{-g} = \frac{2w\sin\theta}{g}.$$

The required relationship between the horizontal distance the jet will travel, (S), and the angle θ of the nozzle, will be

$$S = w\cos\theta \times \frac{2w\sin\theta}{g} = \frac{w^2}{g}(2\sin\theta\cos\theta) = \frac{w^2}{g}\sin 2\theta.$$

The derivative of S with respect to θ will be $\dfrac{dS}{d\theta} = \dfrac{w^2}{g} \times \cos 2\theta \times 2$.

Since w, g, and the factor 2 are all constants, the derivative will be zero when $\cos 2\theta$ is zero, which will be when 2θ is $90°$. Therefore the distance S will be maximum when θ is $45°$, which fortunately verifies the intuitive conclusion.

Table 2-2. Some Common Derivatives

Function F(t)	Derivative
a (constant)	0 (zero)
a t	a
t^n	$n\,t^{(n-1)}$
$\sin \omega t$	$\omega\cos \omega t$
$\cos \omega t$	$-\omega\sin \omega t$
$\tan \omega t$	$\dfrac{\omega}{\cos^2 \omega t}$
$e^{\omega t}$	$\omega e^{\omega t}$
$e^{-\omega t}$	$-\omega e^{-\omega t}$
$\log_e t = |\ln|t$	$\dfrac{1}{t}$
For u and v both functions of t:	
$u + v$	$\dfrac{du}{dt} + \dfrac{dv}{dt}$
$u\,v$	$u\dfrac{dv}{dt} + v\dfrac{du}{dt}$
$\dfrac{u}{v}$	$\dfrac{v\dfrac{du}{dt} - u\dfrac{dv}{dt}}{v^2}$

3
Integral Calculus

Integral calculus has as its basis the mathematical operation of integration, which is generally considered to be the reverse of the operation of taking the derivative of a function. What this means is that in some problems, the derivative dx/dt is known, and the requirement is to determine the original function $f(t)$ for which dx/dt is the derivative. This can often, though not always, be done through integration.

Integration is always performed on the differential of a variable. The quantities dt and dx are the differentials of the variables t and x, respectively. If the relation $x = f(t)$ is given, then the function of t, which is obtained by evaluating the derivative dx/dt, is customarily designated $f'(t)$. That is,

$$x = f(t), \frac{dx}{dt} = f'(t), \frac{d^2x}{dt^2} = f''(t), \text{ and so on.}$$

The relationship between the differentials of x and t, (dx and dt), is consequently

$$dx = f'(t)dt.$$

Differentiation is the process of obtaining the differential of a function. *Integration*, the inverse operation, involves obtaining the original function from the differential. The integration operation is flagged by the $\int$ integration sign.

In mathematical symbology, $dx = f'(t) dt$ identifies the differentiation operation, while $\int f'(t) dt = f(t) = x$ identifies the operation of integration.

Problem Areas

Integration differs from differentiation in one notable respect—while it is always possible to differentiate any function involving the independent variable, it is not possible to integrate all such functions. Certain functions cannot be integrated. This is because although every function has a derivative, not every function is the derivative of some other function.

Unlike differentiation, integration is not a straightforward mechanical procedure. In fact, the basis of performing integration in most cases is having available a table of integration formulas, which has been prepared over time by inverting various formulas for differentiation. For example, the fact that

$$\int \cos x \, dx = \sin x$$

is only known because it is known that

$$\frac{d}{dx}(\sin x) = \cos x.$$

If the function f'(t) dt is to be integrated, the problem is usually one of manipulating the function so that it is compatible with some formula in the table of integrals.

Another complication is that any function that can be integrated will have more than one integral; in fact, it will have many integrals. It can been proven, fortunately, that if two separate functions are both integrals of another function, then these two functions can differ only by a constant. Since it is obligatory to state the solution to a problem of integration in its most general form, it is customary to add on an arbitrary constant to the function obtained by integrating. That is,

$$\int f'(t) dt = f(t) + C.$$

Note that

$$\frac{d}{dt}[f(t) + C] = \frac{d}{dt}f(t) + \frac{d}{dt}C = f'(t)$$

since the derivative of a constant is zero.

In a specific problem, the constant C may have a particular value, which can be determined by applying initial conditions.

Practical Uses of Integration

Integration has practical value in at least three areas.

- If the rate of change of the dependent variable is known, integration over a specified range of the independent variable will yield the cumulative value of the dependent variable over the chosen range.

- An offshoot of this is the ability to calculate the area under a specified section of a curve. This area, divided by the length of the base under the curve, will give the true average value of the dependent variable represented by the curve over the given range.

- The solution of differential equations that describe the dynamic behavior of certain control system components requires the capability to do integration.

A limited list of integration formulas is contained on page 50.

Example 1: Powers and Constants

The formulas in the table on page 50 show that integration is no problem if the expression to be integrated involves only constants and powers of the variable.

Integrate the expression $a + bx + cx^2$, where a, b, and c are constants.

$$\int (a + bx + cx^2)dx = \int adx + \int bxdx + \int cx^2dx$$

$$= a\int dx + b\int xdx + c\int x^2dx = ax + b\frac{x^2}{2} + c\frac{x^3}{3} + C$$

Example 2: $\sin^3 x$

Evaluate $\int \sin^3 xdx$.

This example can clarify two important points regarding the integration process. First, the integration rule which applies to *powers* of the variable involved, *does not apply to powers of functions of that variable.*

Specifically,

$$\int x^3 dx = \frac{x^4}{4} + C, \text{ but } \int \sin^3 xdx \text{ is not } \frac{\sin^4 x}{4} + C.$$

This is a trap into which many students have fallen.

The second point concerns a maneuver that sometimes is needed to effect the integration. It involves forcing part of the function to be integrated past the differential sign, so that instead of integrating with respect to the variable involved, the integration is carried out with respect to a function of that variable. This sounds complicated, but the example will help to simplify the process.

Note that $\int \sin^3 x\, dx = \int \sin^2 x \times \sin x\, dx$. By forcing the function $\sin x$ past the differential sign, $\sin x\, dx$ becomes $d\,(-\cos x)$. The expression inside the brackets is in fact $\int \sin x\, dx$. To confirm that this procedure is mathematically correct,

$$\frac{d}{dx}(-\cos x) = \sin x, \text{ so that } d(-\cos x) = \sin x\, dx.$$

Since $\sin^2 x + \cos^2 x = 1$, then $\sin^2 x = 1 - \cos^2 x$. Therefore,

$$\int \sin^2 x \sin x\, dx \text{ becomes } \int (1 - \cos^2 x)\, d(-\cos x).$$

The integration can now be done on a term by term basis. For convenience, let $\cos x = u$. On substituting, the expression becomes

$$-\int(1 - u^2)\, du = -\int 1\, du + \int u^2\, du = \frac{u^3}{3} - u + C.$$

Therefore,

$$\int \sin^3 x\, dx = \frac{\cos^3 x}{3} - \cos x + C.$$

To verify that the integration has been done correctly, the result should be differentiated.

$$\frac{d}{dx}\left(\frac{\cos^3 x}{3} - \cos x + C\right) = \frac{(3\cos^2 x)(-\sin x)}{3} + \sin x$$

$$= (1 - \sin^2 x)(-\sin x) + \sin x = -\sin x + \sin^3 x + \sin x = \sin^3 x$$

Example 3: cos 2x

This example illustrates still another trap to be avoided. The table of integrals shows that $\int \cos x \, dx = \sin x + C$. From this one could conclude that $\int \cos 2x \, dx = \sin 2x + C$, but this is not correct.

Substitute u=2x. Then

$$x = \frac{u}{2} \text{ and } dx = \frac{1}{2}du$$

$$\int \cos 2x \, dx = \int \cos u \frac{1}{2} du = \frac{1}{2} \int \cos u \, du = \frac{1}{2}\sin u + C = \frac{1}{2}\sin 2x + C.$$

The fact that any differential will be a product of a function f' (x) multiplied by dx, and that these two expressions must be compatible when integration is performed, must be respected. Failure to do this is probably the biggest single cause of mistakes when working out integrals. For the differential term d ( ), what is inside the brackets has to be correct. Cos 2x dx cannot be integrated as it stands. Cos 2x d(2x) is required to compute.

Example 4: Substitution of Variables

Success in integrating a function often depends on an artful substitution of variables, which in turn depends on experience.

Integrate $\int \frac{1}{a-x} dx$ (a is a constant.

Substitute $(a - x) = v$. Then $x = a - v$, $\frac{dx}{dv} = -1$, and $dx = - dv$.

The integral becomes

$$\int \frac{1}{v}(-dv) = -\int \frac{dv}{v} = -\ln|v| + C = C - \ln|a - x|.$$

Example 5: Fractions

If the expression to be integrated is a fraction, consideration should be given to trying to put the entire numerator under the differential sign. The method of solution may then become apparent from the new appearance of the expression. Bear in mind that a constant can be added or subtracted to a function under the differential sign d( ) without altering the mathematical correctness. Specifically, $dx = d (x + a) = d (x - a)$.

Integrate $\int \dfrac{(1+2x)}{2+x+x^2}dx$.

$$\int (1+2x)dx \;=\; x+2\dfrac{x^2}{2}+C \;=\; x+x^2+C\,.$$

The choice of a value for C is arbitrary; hence C can be 2. Therefore,

$$(1+2x)dx \;=\; d(2+x+x^2)\,, \text{ and}$$

$$\int \dfrac{(1+2x)}{2+x+x^2}dx \;=\; \int \dfrac{1}{2+x+x^2}d(2+x+x^2)\,.$$

This is one of the standard forms $\int \dfrac{1}{u}du$, with $u = 2+x+x^2$.

Therefore, the solution is

$$\int \dfrac{(1+2x)}{2+x+x^2}dx \;=\; \ln\left|2+x+x^2\right| + A\,, \; A \text{ being an arbitrary constant.}$$

Example 6: Using Partial Fractions

A fraction in which the numerator and the denominator both consist only of powers of the independent variable, is called a *rational function*. The expression that was integrated in Example 4 is in this category. If the numerator of the rational function is of a lower degree than the denominator, another approach to integrating the expression may be to break the expression into partial fractions.

For example, it is required to integrate $\dfrac{10x+6}{x^2-2x-3}$.

In this case the denominator will factor into $(x-3)$ and $(x+1)$.

$$\dfrac{10x+6}{x^2-2x-3} \;=\; \dfrac{P}{x-3}+\dfrac{Q}{x+1}$$

provided that values for P and Q can be determined. Then,

$$\dfrac{P}{x-3}+\dfrac{Q}{x+1} \;=\; \dfrac{Px+P+Qx-3Q}{x^2-2x-3} \;=\; \dfrac{(P+Q)x+(P-3Q)}{x^2-2x-3}\,.$$

Comparing this numerator with the original numerator $10x + 6$, it is clear that $(P + Q) = 10$, and $(P - 3Q) = 6$. From these two equations, $P = 9$, and $Q = 1$.

Consequently,

$$\int \frac{10x + 6}{x^2 - 2x - 3}dx = \int \frac{9}{x-3}dx + \int \frac{1}{x+1}dx$$

which can be easily integrated.

Example 7: Numerator Higher Order than Denominator

If the numerator of the rational function is of a higher order than the denominator, then the approach is to divide the denominator into the numerator and integrate the results.

Integrate $\dfrac{x^3 + 3x}{x^2 - 2x - 3}$. First, divide the denominator into the numerator.

$$
\begin{array}{r}
x + 2 \\
x^2 - 2x - 3 \overline{\smash{\big)}\ x^3 + 3x} \\
\underline{x^3 - 2x^2 - 3x} \\
2x^2 + 6x \\
\underline{2x^2 - 4x - 6} \\
10x + 6
\end{array}
$$

$\int \dfrac{x^3 + 3x}{x^2 - 2x - 3}dx$ becomes $\int x\,dx + \int 2\,dx + \int \dfrac{10x + 6}{x^2 - 2x - 3}dx$.

Example 8: Changing to an Angular Mode

If an expression contains factors such as x, a (a constant), and $\sqrt{a^2 - x^2}$, it might be useful to switch to a new variable θ, where θ is an acute angle in a right angle triangle for which

$$x, a, \text{ and } \sqrt{a^2 - x^2} \text{ are the sides.}$$

For example, $\int \dfrac{x}{\sqrt{a^2 - x^2}}dx$ is to be determined.

The ratio $\dfrac{x}{\sqrt{a^2-x^2}}$ could be represented as the tangent of the angle θ in

Figure 3-1. Consequently, let $\dfrac{x}{\sqrt{a^2-x^2}} = \tan\theta$.

Then $\dfrac{x}{a} = \sin\theta$, $x = a\sin\theta$, $\dfrac{dx}{d\theta} = a\cos\theta$, and $dx = a\cos\theta d\theta$.

Therefore, $\displaystyle\int\dfrac{x}{\sqrt{a^2-x^2}}dx = \int\tan\theta\times a\cos\theta d\theta = a\int\sin\theta d\theta$.

$$= a(-\cos\theta)+C = a\left(-\dfrac{\sqrt{a^2-x^2}}{a}\right)+C = C-\sqrt{a^2-x^2}.$$

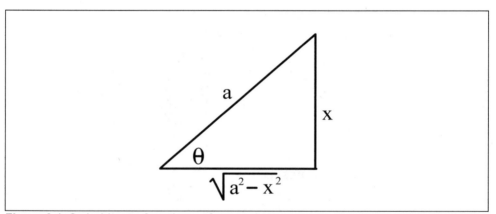

Figure 3-1. Switching to functions of an angle θ.

Example 9: $\sin^2 x$

Evaluate $\displaystyle\int\sin^2 x dx$.

A word of caution here that is worth repeating: The rule that governs integrating powers of a variable does *not* apply to integrating powers of functions of the variable. Specifically, while

$\displaystyle\int x^2 dx$ is equal to $\dfrac{x^3}{3}$, $\displaystyle\int\sin^2 x dx$ is not equal to $\dfrac{\sin^3 x}{3}$.

To integrate the function $\sin^2 x$, the function has to be converted into a more workable form, which turns out to be

$$\sin^2 x = \frac{1 - \cos 2x}{2}.$$

In Chapter 1 on trigonometric functions, the expression for the cosine of a sum was developed.

$$\cos(x + y) = \cos x \cos y - \sin x \sin y.$$

Consequently, $\cos 2x = \cos(x + x) = \cos x \cos x - \sin x \sin x$

$$= \cos^2 x - \sin^2 x.$$

Since $\sin^2 x + \cos^2 x = 1$, $\cos 2x = (1 - \sin^2 x) - \sin^2 x = 1 - 2\sin^2 x$.

Rearranging this, $\sin^2 x = \dfrac{1 - \cos 2x}{2}$.

Thus, $\displaystyle\int \sin^2 x \, dx = \int \left(\frac{1 - \cos 2x}{2}\right) dx = \int \frac{1}{2} dx - \int \frac{1}{2}\cos 2x \, dx$.

$$\int \frac{1}{2} dx = \frac{1}{2}\int dx = \frac{x}{2}$$

For $\displaystyle\int \frac{1}{2}\cos 2x \, dx$, substitute $u = 2x$. Then, $\dfrac{du}{dx} = 2$, and $dx = \dfrac{du}{2}$.

$$\int \frac{1}{2}\cos 2x \, dx = \frac{1}{2}\int \cos u \frac{du}{2} = \frac{1}{4}\int \cos u \, du = \frac{1}{4}\sin u = \frac{1}{4}\sin 2x$$

Finally, $\displaystyle\int \sin^2 x \, dx = \frac{x}{2} - \frac{1}{4}\sin 2x + C$.

This result should be verified by taking the derivative with respect to x.

$$\frac{d}{dx}\left(\frac{x}{2} - \frac{\sin 2x}{4}\right) = \frac{1}{2} - \frac{\cos 2x}{4} \times 2 = \frac{1}{2}(1 - \cos 2x) = \sin^2 x$$

Example 10: Square Root of ($a^2 - x^2$)

Sometimes the solution for one integral will provide a means of solving another. The integral

$$\int \sqrt{a^2 - x^2} \, dx$$

looks as though it should be fairly simple to evaluate but proves to be otherwise. The approach is to convert to the trigonometric mode.

Figure 3-2 is a right angled triangle with its hypotenuse equal to a, and the side opposite the angle θ equal to x.

This makes the remaining side $\sqrt{a^2 - x^2}$.

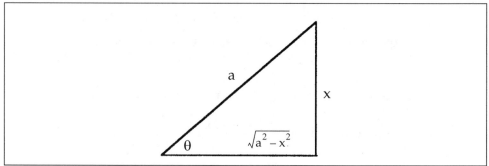

Figure 3-2. The trigonometric mode.

In this triangle,

$$\frac{\sqrt{a^2 - x^2}}{a} = \cos\theta, \text{ and } \sqrt{a^2 - x^2} = a\cos\theta.$$

$$\frac{x}{a} = \sin\theta, x = a\sin\theta, \text{ and } dx = a\cos\theta d\theta$$

$$\therefore \int\sqrt{a^2 - x^2}dx = \int a\cos\theta . a\cos\theta d\theta = a^2\int\cos^2\theta d\theta$$

$$= a^2\left[\int(1 - \sin^2\theta)d\theta\right] = a^2\left[\int 1 d\theta - \int\sin^2 d\theta\right]$$

$$= a^2\left[\theta - \left(\frac{\theta}{2} - \frac{1}{4}\sin 2\theta\right)\right] \quad \text{(from Example 9)}$$

$$= a^2\left(\frac{\theta}{2} + \frac{1}{4}\sin 2\theta\right)$$

However, the solution should be in terms of x and a, not θ and a.

For the first term in the brackets, $\theta = \sin^{-1}\frac{x}{a}$

$$\frac{1}{4}\sin 2\theta = \frac{1}{4}(2\sin\theta\cos\theta) = \frac{1}{2}\times\frac{x}{a}\times\frac{\sqrt{a^2 - x^2}}{a} \quad \text{(from Figure 3-2)}$$

$$\therefore \int \sqrt{a^2 - x^2}\, dx = a^2 \left(\frac{\sin^{-1}\frac{x}{a}}{2} + \frac{1}{2a^2}x\sqrt{a^2 - x^2} \right)$$

$$= \frac{a^2}{2}\sin^{-1}\frac{x}{a} + \frac{x}{2}\sqrt{a^2 - x^2} + C = \frac{1}{2}\left(a^2\sin^{-1}\frac{x}{a} + x\sqrt{a^2 - x^2}\right) + C.$$

Integration Over a Specified Range

So far, the integration process has resulted in integrals that are general in nature, in the sense that after integration has been performed, it is necessary to add a constant to the integral. This recognizes the fact that if two functions of the same independent variable differ only by a constant, then they will have the same derivative.

In some problems, however, the requirement is to integrate the expression over a specified range of the independent variable. In these cases, the constant does not apply, but it is necessary to identify what the upper and lower limits of the integration are to be. These limits are specified mathematically by placing them at the top and bottom of the integral sign. After the integral of the function has been determined, the final result will be the value of the integral at the upper limit of the independent variable, minus the value of the integral at the lower limit of the independent variable. The example below should help to illustrate this.

Example 11: Tank Filling Case

Figure 3-3 shows a 200 litre tank, which is being filled with a liquid by gravity from a large reservoir. The reservoir is large enough that drawing off 200 litre will not appreciably change its level. The driving force that moves the liquid is proportional to the difference in the levels of the tank and the reservoir. When the valve is opened, the initial flow rate is 2 l/s into the tank, but as the tank fills, this difference decreases, the driving force diminishes, and the rate of flow falls off.

Figure 3-4 shows how the flow rate decreases over time, starting at 2 l/s at t = 0. The relationship shown by the graph is typical of situations in which the driving force that creates the change in the dependent variable is in proportion to the difference between the instantaneous value of the dependent variable and its ultimate value over time. As the dependent variable approaches its ultimate value, in this case zero flow, the driving force falls off, as does the rate of change of the dependent variable.

If x is the liquid flow rate, the equation of the graph will be $x = 2.0e^{-.005t}$.

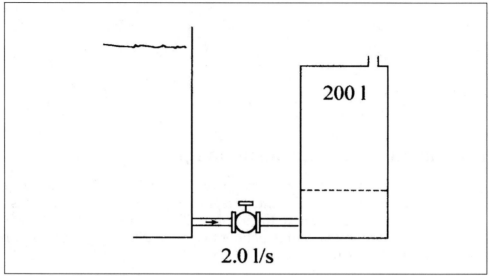

Figure 3-3. Tank Filling.

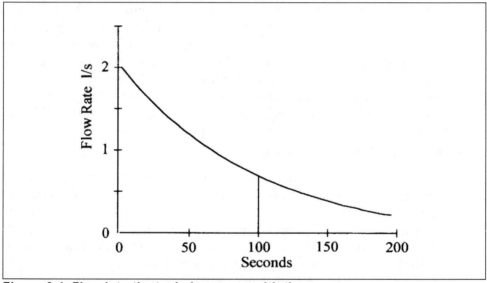

Figure 3-4. Flow into the tank decreases with time.

In this relation, x is in litres per second and t is in seconds. The factor 0.005 is set by the resistance to flow of the valve and piping and the cross sectional area of the tank.

If the flow rate into the tank were maintained at the initial rate of 2 l/s, then the tank would fill up in 100 s. A logical question might be how much liquid will actually be in the tank after 100 s, considering that the flow rate decreases as time goes on? To determine this, it is necessary to integrate the function x = f (t) over the range t = 0 to t = 100. The cumulative flow value is actually represented graphically by the area under the curve from t = 0 to t = 100. It is this area value that the integration will hopefully reveal.

$$\int_0^{100} f(t)dt = \int_0^{100} 2e^{-.005t}dt = 2\left[\frac{1}{-.005}e^{-.005t}\right]_0^{100} = -400[e^{-.005t}]^{100}$$

$$-400(0.607 - 1) = -400 \times -.393 = 157 \text{ litres}.$$

The average flow rate over this time period would be 157 *l* in 100 s or 1.57 *l*/s.

The solution to this example is fraught with mathematical shorthand. To avoid possible misunderstanding,

$[e^{-.005t}]_0^{100}$ means the value of $e^{-.005t}$ when *t* = 100, minus the value of

$e^{-.005t}$ when *t* = 0.

The next set of examples illustrates the principle of arriving at the desired solution by first setting up a differential increment of the entity to be determined, then integrating that increment over the relevant range to obtain the complete entity.

Example 12: Area of a Circle

The circumference of a circle is equal to $2\pi \times$ the radius of the circle. Starting with this basic fact, the area of a circle can be determined by integration without much difficulty.

Figure 3-5 shows a circle with a radius r. Inside this circle is an elemental ring of width dx at a distance x from the center of the circle. The area inside the elemental ring will be the product of its length and width, that is, 2π x times dx. If this expression can be integrated from x = 0 to x = r, this will determine the area of the circle itself.

$$\text{Area} = \int_0^r 2\pi x dx = 2\pi \int_0^r x dx = 2\pi \left[\frac{x^2}{2}\right]_0^r = 2\pi \left[\frac{r^2}{2} - 0\right] = \pi r^2$$

This result, of course, is a revelation to no one.

Example 13: Surface Area of a Sphere

A similar procedure can be used to determine the surface area of a sphere.

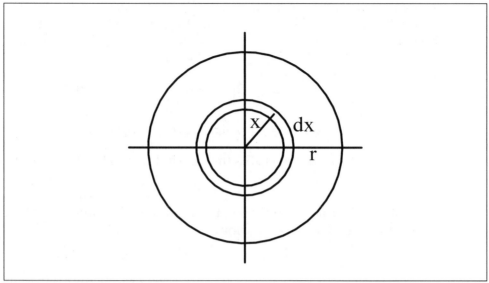

Figure 3-5. Area of a circle.

Figure 3-6 shows a sphere with a radius r. Distance on the surface of the sphere is represented by the variable a. In this case, a = arc PB. The elemental surface is a ring around the outside of the sphere. Its width will be da. Its length will be 2π times the distance AP.

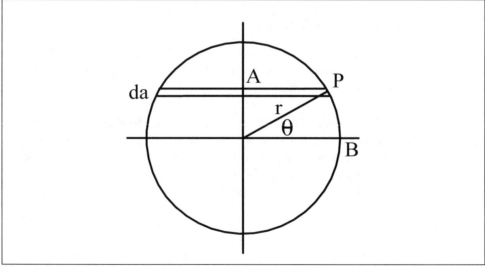

Figure 3-6. Area of a sphere.

If the radius vector makes an angle θ with the horizontal axis, then for an incremental change $d\theta$ in θ,

$$da = r \times d\theta.$$

Also, $\dfrac{AP}{r} = \cos\theta$, and $AP = r\cos\theta$.

Consequently, the expression to be integrated is $2\pi \times r \cos\theta \times r\, d\theta$. The expression should be integrated between the limits $\theta = -90$ degrees to $\theta = +90$ degrees. In radians, this means $-\pi/2$ to $+\pi/2$.

$$\text{Area} = \int_{-\frac{\pi}{2}}^{\frac{\pi}{2}} 2\pi r^2 \cos\theta\, d\theta = 2\pi r^2 \int_{-\frac{\pi}{2}}^{\frac{\pi}{2}} \cos\theta\, d\theta = 2\pi r^2 \left[\sin\theta\right]_{-\frac{\pi}{2}}^{\frac{\pi}{2}}$$

$$= 2\pi r^2 [1 - (-1)] = 2\pi r^2 \times 2 = 4\pi r^2$$

Example 14: Volume of a Sphere

The integration process can also be used to determine the volume of a sphere.

Figure 3-7 shows a sphere with a radius r. The incremental element is a disc cut through the sphere at distance y from the origin. The radius of the disc is the distance AB, and its thickness is dy.

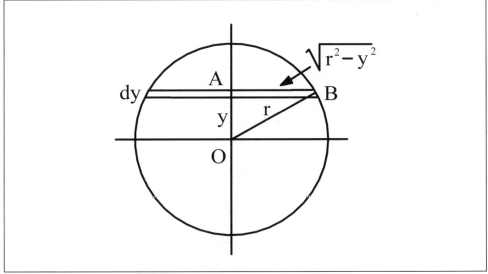

Figure 3-7. Volume of a sphere.

Since the triangle OAB is right angled,

$$(AB)^2 + y^2 = r^2, \text{ and } (AB)^2 = r^2 - y^2.$$

The volume of the elemental disc will be $\pi\,(AB)^2 \times dy = \pi\,(r^2 - y^2)\,dy$.

To arrive at the volume of the sphere, this expression must be integrated from the bottom of the sphere to its top, which means from $-r$ to $+r$.

$$\int_{-r}^{r} \pi(r^2 - y^2)dy = \pi r^2 \int_{-r}^{r} dy - \pi \int_{-r}^{r} y^2 dy = \pi r^2 [y]_{-r}^{r} - \pi \left[\frac{y^3}{3}\right]_{-r}^{r}$$

$$= \pi r^2 [r - (-r)] - \frac{\pi}{3}[r^3 - (-r^3)] = \pi r^2(2r) - \frac{\pi}{3}(2r^3) = 2\pi r^3 - \frac{2}{3}\pi r^3$$

$$= \frac{4}{3}\pi r^3.$$

Example 15: The Escape Velocity

Determine the vertical velocity that a mass m would have to be given at the surface of the earth to break free from the gravitational pull of the earth and remain in orbit. This velocity is sometimes called the escape velocity.

It is a fact of physics that work and energy are interchangeable. Accordingly, the approach to this problem is to calculate the work that would have to be done in moving the mass from the earth's surface out into space where the gravitational pull of the earth is no longer effective. Then the kinetic energy that the mass must have as it starts out must equal the required amount of work. If the kinetic energy is known, then the velocity can be calculated.

The relationship governing the gravitational force F between 2 masses is

$$F = k\,\frac{m_1 m_2}{d^2}$$

where

m_1 and m_2 are the masses of the mutually attracted objects, in kilograms. If m_1 is the mass of the earth, then
$m_1 = 5.983 \times 10^{24}$ kg.

d is the distance between the centers of gravity of the objects, in metres.

k is the gravitational constant (6.670×10^{-11}).

With k at this value, the m's in kilograms, and d in metres, F will be in Newtons.

Let the vertical elevation of the object be y, measured from the center of the earth. This puts the starting point of the object at the radius r of the earth, or $y = 6.371 \times 10^6$ m.

For a small change in elevation dy, the work done will be the force times the distance.

$$dw = k\frac{m_1 m}{y^2}dy$$

To find the total work required, this function should be integrated from y = earth's radius ($r = 6.371 \times 10^6$ metres) to y = infinity.

$$\int dw = \int_r^\infty k\frac{m_1 m}{y^2}dy = km_1 m\int_r^\infty \frac{1}{y^2}dy = km_1 m\int_r^\infty y^{-2}dy$$

$$= km_1 m\left[-\frac{1}{y}\right]_{6.731 \times 10^6}^\infty = km_1 m\left(-0 - \left(-\frac{1}{6.731 \times 10^6}\right)\right)$$

$$= (6.670 \times 10^{-11}) \times (5.983 \times 10^{24}) \times m \times \left\{\frac{1}{6.371 \times 10^6}\right\}$$

$$= 6.264 \times 10^7 \times m$$

Therefore, the kinetic energy that must be given to the object will be

$\frac{1}{2}mv^2 = 6.264 \times 10^7 \times m$, from which $v^2 = 1.2527 \times 10^8$, and

$v = 1.119 \times 10^4$ m/s, or 6.95 miles/s.

Example 16: Area of a Segment of a Circle

The problem in this example is to develop a formula for the area of a segment of a circle, as shown in Figure 3–8.

In the circle shown, the segment of concern is that bounded by the chord AD, which is a distance d from the center of the circle, and the arc ACD. The formula that is derived should be in terms of the radius r of the circle and the distance d.

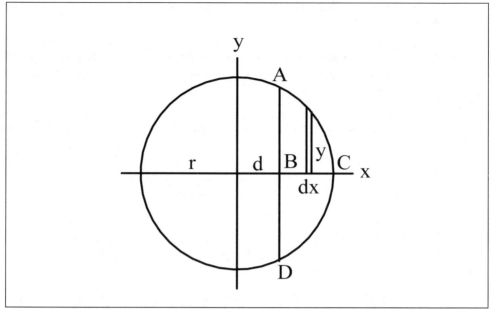

Figure 3-8. Area of a segment of a circle.

From the figure, it is apparent that the area of the segment will be twice the area of the half segment described by the corners A, B, and C. Within that half segment is an elemental strip that is dx in width and y in height, so that its area is y dx. The task, therefore, will be to integrate the expression 2 y dx from x = d to x = r.

The equation for the circle is $x^2 + y^2 = r^2$. Therefore, $y = \sqrt{r^2 - x^2}$.

The integral to be computed, then, is $2\int_d^r \sqrt{r^2 - x^2}\,dx$.

Example 10 showed that

$$\int \sqrt{r^2 - x^2}\,dx = \frac{1}{2}\left(r^2 \sin^{-1}\frac{x}{r} + x\sqrt{r^2 - x^2}\right)$$

$$\therefore 2\int_d^r \sqrt{r^2 - x^2}\,dx = 2\left[\frac{1}{2}\left(r^2 \sin^{-1}\frac{x}{r} + x\sqrt{r^2 - x^2}\right)\right]_d^r$$

$$= 2\left[\frac{1}{2}\left(r^2 \sin^{-1}\frac{r}{r} + r\sqrt{r^2 - r^2}\right) - \frac{1}{2}\left(r^2 \sin^{-1}\frac{d}{r} + d\sqrt{r^2 - d^2}\right)\right]$$

$\sin^{-1}\frac{r}{r} = $ (angle whose sine is 1) $= 90$ deg, or $\frac{\pi}{2}$ in radians.

$$\therefore \text{Area of the segment} = \frac{\pi}{2}r^2 - \left(r^2\sin^{-1}\frac{d}{r} + d\sqrt{r^2 - d^2}\right)$$

This equation looks weird for an area expression, but it can be tested.

If d = 0, the chord that forms the segment will lie on the vertical (y) axis, and the segment will occupy half of the circle. With d = 0,

$$\sin^{-1}\frac{d}{r} = \sin^{-1}0 = 0, \text{ as does } d\sqrt{r^2 - d^2}, \text{ and the area of the segment}$$

$$= \frac{\pi r^2}{2}.$$

If d = r, the area should be zero. For d = r,

$$\sin^{-1}\frac{r}{r} = \sin^{-1}1 = \frac{\pi}{2} \text{ radians, and } r\sqrt{r^2 - r^2} = 0$$

The area of the segment becomes $\frac{\pi r^2}{2} - r^2\frac{\pi}{2} = 0$, which computes.

Table of Basic Integrals

In this table, a, n, and C are constants, and x, u, and v are functions of t.

1.   The integral of a sum is the sum of the integrals.

$$\int (dx + du + dv) = \int dx + \int du + \int dv$$

2.   A constant multiplying the function to be integrated can be moved outside of the integral sign.

$$\int a\, dx = a \int dx$$

3. $\int x^n dx = \dfrac{x^{n+1}}{n+1} + C$, provided that n is not –1.

4. $\int x^{-1} dx = \int \dfrac{1}{x} dx = \ln|x| + C$ (Note 1)

5. $\int \cos x\, dx = \sin x + C$

6. $\int \sin x\, dx = -\cos x + C$

7. $\int \tan x\, dx = -\ln|\cos x| + C$ (Note 1)

8. $\int e^{ax} dx = \dfrac{1}{a} e^{ax} + C$

9. $\int \sin^2 x\, dx = \dfrac{x}{2} - \dfrac{1}{4}\sin 2x + C$

10. $\int \cos^2 x\, dx = \dfrac{x}{2} + \dfrac{1}{4}\sin 2x + C$

11. $\int \sqrt{a^2 - x^2}\, dx = \dfrac{1}{2}\left(a^2 \sin^{-1}\dfrac{x}{a} + x\sqrt{a^2 - x^2}\right) + C$

Note 1: In mathematical notation, the natural logarithm of a number x, that is, its logarithm to the base e, is written ln |x|.

4

Infinite Series

It sometimes happens that a function f(t) of the variable t appears as the sum of a number of terms, each of which in itself is a function of t. One such example is

$$f(t) = \sin \omega t + 2 \sin (\omega t)^2 + 3 \sin (\omega t)^3 + \text{etc.}$$

This type of function can become useful under the following circumstances.

- There is no limit to the number of terms.
- The format of the terms follows a discernible pattern.

In other words, if n designates the number of an individual term in the function (n = 1, 2, 3, 4, etc.), then a formula for the general, or n^{th}, term can be written. In the example above, the nth term is $n \sin (\omega t)^n$.

If these two conditions are met, the function f(t) is called an *infinite series*.

An infinite series will often prove productive if it is the type for which the sum of the terms never exceeds a certain finite limit, no matter how many terms are added on. A series of this type is said to be *convergent*. An infinite series whose sum eventually becomes infinite as more and more terms are added is called *divergent*.

For example, the infinite series

$$f(t) = 1 + t + t^2 + t^3 + t^4 + \ldots + t^n + \ldots$$

may be convergent or divergent, depending on the value of t. If t = 1, then the sum goes to infinity as the number of terms becomes larger. If t < 1, however, the sum will be limited to a definite finite number, no matter how many terms are included. If t = 1/2, for instance, the sum will never be greater than 2.

An allied characteristic of a series that is convergent is that as the number of terms (n) gets larger, the value of the nth term approaches zero. This is illustrated for the series above in Figure 4-1, with t = ½ and t = 2. When the nth term is becoming larger as n increases, it is an indication that the sum of the series is going to infinity.

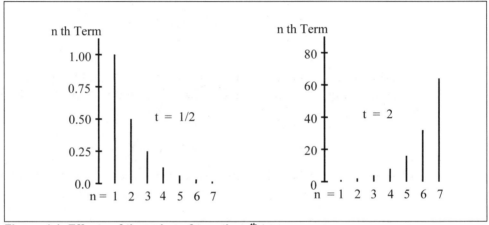

Figure 4-1. Effects of the value of t on the nth term.

In the graphs in Figure 4-1, it should be noted that a curve joining the tips of the vertical bars representing the values of the terms has no significance. This is because n is an integer. No intermediate values lie between the whole numbers.

Power Series

A power series is a special form of infinite series, in which the terms are ascending powers of the variable, multiplied by a constant coefficient. The general form of a power series is

$$f(t) = a_0 + a_1 t + a_2 t^2 + a_3 t^3 + \ldots + a_n t^n + etc.$$

The nth Term

If the first few terms of an infinite series are given, it may be possible to develop the formula for the nth term. There is at least one reason for doing this, namely, the test for the convergence of the series (described later) requires the expression for the nth term.

Usually the expression (formula) can be deduced from inspecting the terms progressively. Arranging the various factors in a table will be helpful. In a power series, it is necessary to establish three things: (1) the power, (2) the coefficient, and (3) the sign, in order to define the nth term completely. An example will help to illustrate the method.

Evaluate the nth term of the series

$$f(t) = t - \frac{t^3}{3!} + \frac{t^5}{5!} - \frac{t^7}{7} + \frac{t^9}{9} - etc.$$

The exclamation mark (!) here is the factorial symbol.

The factors tabulate as follows:

Table 4-1. n^{th} Term Factors

Term	Power of t	Coefficient	Sign
1	1	1	+
2	3	1/3!	−
3	5	1/5!	+
4	7	1/7!	−
5	9	1/9!	+
n	2n − 1	$\dfrac{1}{(2n-1)!}$	$(-1)^{(n+1)}$

The nth term of this series is consequently

$$\left(-1\right)^{(n+1)} \frac{1}{(2n-1)!} t^{(2n-1)}.$$

Test for Convergence

There are three established tests for convergence of an infinite series. Only one of these, the *ratio test*, will be mentioned here. It is based on determining the ratio of the $(n + 1)^{th}$ term to the nth term. These two terms are, of course, consecutive terms. The ratio test is prescribed in the following way.

An infinite series $f(t) = u_1 + u_2 + u_3 + \ldots$ is convergent if the ratio

$$\frac{u_{(n+1)}}{u_n}$$

is numerically <1 as n approaches infinity, and divergent if the ratio is > or = 1 as n approaches infinity.

Example 1: Convergence Test 1

Test the series $f(t) = 1 + \frac{1}{2} + \frac{1}{4} + \frac{1}{8}$ + etc. for convergence.

The nth term in this series is

$$\frac{1}{2^{(n-1)}}.$$

Substituting (n + 1) for n in the expression for the nth term, gives

$$\frac{1}{2^n}$$

which is the $(n + 1)^{th}$ term. The ratio of the $(n + 1)^{th}$ term to the nth term is

$$\frac{\dfrac{1}{2^n}}{\dfrac{1}{2^{(n-1)}}}$$

which simplifies to $\frac{1}{2}$. Since the result turns out to be <1, the series converges.

Example 2: Convergence Test 2

Taking the general case of Example 1, for the series

$$f(t) = 1 + t + t^2 + t^3 + t^4 + \text{etc.},$$

the nth term is $t^{(n-1)}$, and the $(n + 1)^{th}$ term is t^n. The required ratio is

$$\frac{t^n}{t^{(n-1)}} = t.$$

This is <1 for values of t < 1. Therefore, if t < 1, the series converges. If t > 1, it diverges.

Example 3: Convergence Test 3

Test the series $f(t) = t - \dfrac{t^3}{3!} + \dfrac{t^5}{5!} - \dfrac{t^7}{7!} + \dfrac{t^9}{9!} -$ etc. for convergence.

As was previously deduced, the nth term of this series is

$$(-1)^{(n+1)} \frac{t^{(2n-1)}}{(2n-1)!}.$$

The $(n+1)^{th}$ term, which is obtained by replacing n with $(n+1)$, is consequently,

$$(-1)^{(n+2)} \frac{t^{(2n+1)}}{(2n+1)!}.$$

When determining the ratio, the magnitude only is relevant. The sign has no influence on whether or not the series converges.

The ratio of consecutive terms is $\dfrac{t^{(2n+1)}}{(2n+1)!} \div \dfrac{t^{(2n-1)}}{(2n-1)!}$

$$\frac{(2n-1)!}{(2n+1)!} \times \frac{t^{(2n+1)}}{t^{(2n-1)}} = \frac{t^2}{(2n+1) \times 2n}.$$

As n approaches infinity, this ratio approaches zero, a quantity obviously <1, provided that t is not infinite. Therefore the series

$$f(t) = t - \frac{t^3}{3!} + \frac{t^5}{5!} - \frac{t^7}{7!} + \dots \text{ is convergent for all finite values of } t.$$

When determining the $(n+1)^{th}$ term, it is important to realize that the correct procedure is to substitute "$(n+1)$" for "n" in the expression for the nth term. This is not the same thing as adding 1 to the expression that involves the factor n. The fact that this second procedure sometimes yields the same answer is deceiving.

Example 4: The (n+1)th Term

It was previously determined that the nth term of the series

$$f(t) = t - \frac{t^3}{3!} + \frac{t^5}{5!} - \frac{t^7}{7!} + \dots \text{ is } (-1)^{(n+1)} \frac{1}{(2n-1)!} t^{(2n-1)}.$$

To obtain the $(n + 1)^{th}$ term, n is replaced by $(n + 1)$ at each place where n appears. That is, the $(n + 1)^{th}$ term will be

$$(-1)^{[(n+1)+1]} \frac{1}{[2(n+1)-1]!} t^{[2(n+1)-1]} = (-1)^{(n+2)} \frac{1}{(2n+1)!} t^{(2n+1)}.$$

Notice that this is not the same as adding 1 in each expression in which n appears. The $(n + 1)^{th}$ term *is not*

$$(-1)^{(n+1+1)} \frac{1}{(2n-1+1)!} t^{(2n-1+1)} = (-1)^{(n+2)} \frac{1}{2n!} t^{2n}.$$

Maclaurin's Series

Some mathematical functions can be represented by an infinite series, provided that the function is to be evaluated for values of the variable quantity in the series which will cause the series to converge. In the language of mathematicians, this is referred to as being "within the region of convergence" of the series.

Suppose that $f(t) = a_0 + a_1 t + a_2 t^2 + a_3 t^3 + ...$, and that the values of t are restricted to those that cause the series to converge. The problem is to find the values of the various constant coefficients $a_0, a_1, a_2, a_3, ...$. Since the function $f(t) = a_0 + a_1 t + a_2 t^2 + a_3 t^3 ...$ holds for any value of t that produces convergence, the coefficients can be determined by taking successive derivatives with respect to t for both sides of the equation, and then substituting $t = 0$ in the result. ($t = 0$ is a value within the region of convergence.)

This process can be illustrated as follows.

Suppose that $f(t)$ is $\sin t = a_0 + a_1 t + a_2 t^2 + a_3 t^3 + a_4 t^4 + ...$ Taking the derivative with respect to t for both side results in:

$$\frac{d}{dt} = \cos t = a_1 + 2a_2 t + 3a_3 t^2 + 4a_4 t^3 + 5a_5 t^4 + ...$$

$$\frac{d^2}{dt^2} = -\sin t = 1 \times 2a_2 + 2 \times 3a_3 t + 3 \times 4a_4 t^2 + 4 \times 5a_5 t^3 \ ...$$

$$\frac{d^3}{dt^3} = -\cos t = 1 \times 2 \times 3a_3 + 2 \times 3 \times 4a_4 t + 3 \times 4 \times 5a_5 t^2 + ...$$

$$\frac{d^4}{dt^4} = \sin t = 1\times2\times3\times4\,a_4 + 2\times3\times4\times5\,a_5\,t + \dots$$

$$\frac{d^5}{dt^5} = \cos t = 1\times2\times3\times4\times5\,a_5 \dots$$

In each of these equations in t, substitute t = 0.

For f(t), sin (0) = 0 = a_O, and all of the terms following a_O will be zero because they contain t to some power. Thus a_O = 0.

For $\frac{d}{dt}$; $\cos(0) = 1 = a_1$. ∴ $a_1 = 1$.

For $\frac{d^2}{dt^2}$; $-\sin(0) = 0 = 2! \times a_2$. ∴ $a_2 = 0$.

For $\frac{d^3}{dt^3}$; $-\cos(0) = -1 = 3! \times a_3$. ∴ $a_3 = -\frac{1}{3!}$.

For $\frac{d^4}{dt^4}$; $\sin(0) = 0 = 4! \times a_4$. ∴ $a_4 = 0$.

For $\frac{d^5}{dt^5}$; $\cos(0) = 1 = 5! \times a_5$. ∴ $a_5 = +\frac{1}{5!}$.

From the pattern of the values of the coefficients, the conclusion is that

$$\sin t = t - \frac{1}{3!}t^3 + \frac{1}{5!}t^5 - \frac{1}{7!}t^7 + \dots$$

The general case can now be developed. Designate f(t) = F.

$$F = a_0 + a_1 t + a_2 t^2 + a_3 t^3 + a_4 t^4 + a_5 t^5 + \dots$$

$$\frac{dF}{dt} = a_1 + 2a_2 t + 3a_3 t^2 + 4a_4 t^3 + 5a_5 t^4 + \dots$$

$$\frac{d^2 F}{dt^2} = 1 \times 2 \times a_2 + 2 \times 3 \times a_3 t + 3 \times 4 \times a_4 t^2 + 4 \times 5 \times a_5 t^3 + \dots$$

$$\frac{d^3F}{dt^3} = 1 \times 2 \times 3 \times a_3 + 2 \times 3 \times 4 \times a_4 t + 3 \times 4 \times 5 \times a_5 t^2 + \ldots$$

$$\frac{d^4F}{dt^4} = 1 \times 2 \times 3 \times 4 \times a_4 + 2 \times 3 \times 4 \times 5 \times a_5 t + \ldots$$

Substituting t = 0 in each of these relations, results in

$$F(0) = a_O \text{ and } a_O = F(0)$$

$$\frac{dF}{dt}(0) = a_1 \text{ and } a_1 = \frac{dF}{dt}(0)$$

$$\frac{d^2F}{dt^2}(0) = 1 \times 2 \times a_2 \text{ and } a_2 = \frac{1}{2!}\frac{d^2F}{dt^2}(0)$$

$$\frac{d^3F}{dt^3}(0) = 1 \times 2 \times 3 \times a_3 \text{ and } a_3 = \frac{1}{3!}\frac{d^3F}{dt^3}(0)$$

$$\frac{d^4F}{dt^4}(0) = 1 \times 2 \times 3 \times 4 \times a_4 \text{ and } a_4 = \frac{1}{4!}\frac{d^4F}{dt^4}(0).$$

The general expression for f(t) (= F) is consequently

$$F = F(0) + \frac{dF}{dt}(0)t + \frac{1}{2!}\frac{d^2F}{dt^2}(0)t^2 + \frac{1}{3!}\frac{d^3F}{dt^3}(0)t^3 + \frac{1}{4!}\frac{d^4F}{dt^4}(0)t^4 + \frac{1}{5!}\frac{d^5F}{dt^5}(0)t^5 + \ldots$$

This expansion is called Maclaurin's series.

Example 5: $e^{\omega t}$

Expand $f(t) = e^{\omega t}$ and determine the region of convergence.

For convenience, designate f(t) = F.

$$F = e^{\omega t} \text{ and } F(0) = 1$$

$$\frac{dF}{dt} = \omega e^{\omega t} \text{ and } \frac{dF}{dt}(0) = \omega$$

$$\frac{d^2F}{dt^2} = \omega^2 e^{\omega t} \text{ and } \frac{d^2F}{dt^2}(0) = \omega^2$$

$$\frac{d^3F}{dt^3} = \omega^3 e^{\omega t} \text{ and } \frac{d^3F}{dt^3}(0) = \omega^3$$

$$\frac{d^4F}{dF^4} = \omega^4 e^{\omega t} \quad \text{and} \quad \frac{d^4F}{dF^4}(0) = \omega^4$$

Accordingly,

$$e^{\omega t} = 1 + \omega t + \frac{1}{2!}\omega^2 t^2 + \frac{1}{3!}\omega^3 t^3 + \frac{1}{4!}\omega^4 t^4 + \dots$$

The region of convergence can be determined from the table of successive terms, and the form of the nth term.

Term	Power	Coefficient	Sign
1	0	1	+
2	1	ω	+
3	2	$\frac{1}{2!}\omega^2$	+
4	3	$\frac{1}{3!}\omega^3$	+
5	4	$\frac{1}{4!}\omega^4$	+
n	$(n-1)$	$\frac{1}{(n-1)!}\omega^{(n-1))}$	+

The nth term is consequently

$$+ \frac{\omega^{(n-1)}}{(n-1)!}t^{(n-1)}$$

The $(n+1)^{th}$ term will be

$$+ \frac{\omega^{[(n+1)-1]}}{[(n+1)-1]!}t^{(n+1)-1} = + \frac{1}{n!}\omega^n t^n.$$

The ratio of consecutive terms is

$$\frac{\omega^n t^n}{n!} \div \frac{\omega^{(n-1)}t^{(n-1)}}{(n-1)!} = \frac{(n-1)!}{n!} \times \frac{\omega^n t^n}{\omega^{(n-1)}t^{(n-1)}} = \frac{1}{n}\omega t.$$

As n approaches infinity, the limit of

$$\frac{\omega t}{n}$$

approaches zero for all finite values of ω and t. Therefore, the expansion for $e^{\omega t}$ is valid for all finite values of ω and t.

Practical Disadvantage of Maclaurin's Series

The value of the function f(t) can be evaluated by expanding it in a Maclaurin's series provided that the value of t is fairly small so that the t^2, t^3, t^4, etc. terms diminish rapidly. Thus, the value for f(t) would be obtainable to the desired degree of accuracy without having to calculate very many terms. On the other hand, if t were close to the maximum value it could assume without causing the series to diverge, then the values of a great many terms would have to be calculated to obtain the value of f(t) to the desired accuracy.

Taylor's Series

To get around the problem with the Maclaurin's series, one approach could be to develop a different series, not in powers of t, but as a series of powers of (t – a), where a is a constant. That is,

$$f(t) = F = a_0 + a_1 (t-a) + a_2 (t-a)^2 + a_3 (t-a)^3 + a_4 (t-a)^4 + \ldots$$

Successively taking the derivative with respect to t, gives

$$\frac{dF}{dt} = a_1 + 2a_2(t-a) + 3a_3(t-a)^2 + 4a_4(t-a)^3 + \ldots$$

$$\frac{d^2F}{dt^2} = 1\times2\times a_2 + 2\times3\times a_3(t-a) + 3\times4\times a_4(t-a)^2 + \ldots$$

$$\frac{d^3F}{dt^3} = 1\times2\times3\times a_3 + 2\times3\times4\times a_4(t-a) + \ldots$$

$$\frac{d^4F}{dt^4} = 1\times2\times3\times4\times a_4 + 2\times3\times4\times5\times a_5(t-a) + \ldots$$

Setting t = a in all of these expressions, so that (t – a) = 0 gives,

$$F(a) = a_0 \text{ and } a_0 = F(a)$$

$$\frac{dF}{dt}(a) = a_1 \text{ and } a_1 = \frac{dF}{dt}(a)$$

$$\frac{d^2F}{dt^2}(a) = 1 \times 2 \times a_2 \text{ and } a_2 = \frac{1}{2!} \frac{d^2F}{dt^2}(a)$$

$$\frac{d^3F}{dt^3}(a) = 1 \times 2 \times 3 \times a_3 \text{ and } a_3 = \frac{1}{3!} \frac{d^3F}{dt^3}(a)$$

$$\frac{d^4F}{dt^4}(a) = 1 \times 2 \times 3 \times 4 \times a_4 \text{ and } a_4 = \frac{1}{4!} \frac{d^4F}{dt^4}(a).$$

Therefore,

$$f(t) = F = F(a) + \frac{dF}{dt}(a)(t-a) + \frac{1}{2!}\frac{d^2F}{dt^2}(a)(t-a)^2 + \frac{1}{3!}\frac{d^3F}{dt^3}(a)(t-a)^3 + ...$$

In this expression, $F(a)$ is the value of $f(t)$ when $t = a$. $\frac{dF}{dt}(a)$ is the value of the first derivative of $f(t)$ when $t = a$. The higher derivatives follow in the same manner.

This expansion is Taylor's series. To apply it, it is necessary to know the value of $f(t)$ and all of its relevant derivatives at $t = a$. Furthermore, if $t = b$ is the value at which $f(t)$ is to be evaluated, then the value of a should be selected as close as possible to the value of b so that the $(b - a)^2$, $(b - a)^3$, etc. terms will diminish quickly, and fewer terms will be required to calculate the value of $f(t)$ to the desired accuracy.

The Taylor series is an expansion of the function $f(t)$ using the value $t = a$ as the origin or starting point, rather than zero. Comparing the Taylor series with the Maclaurin series shows that the Maclaurin series is the particular case of the Taylor series in which a = zero.

Example 6: Value of a Sine Function Using Taylor's Series

Determine the value of $\sin 32°$ to three decimal places. This could best be done by developing a Taylor series using $a = 30°$ as the origin, since this value is close to the desired value, and the values of $\sin 30°$ and $\cos 30°$ are known to be

$$\frac{1}{2} \text{ and } \frac{\sqrt{3}}{2}, \text{ respectively.}$$

For the calculations to be correct, the values of the angles involved must be expressed in their fundamental units of radians, rather than degrees. Thus $a = 30° = 0.524$ rad. For $f(t) = F = \sin t$,

$$F(0.524) = \sin 0.524 = 0.500$$

$$\frac{dF}{dt}(0.524) = \cos 0.524 = 0.866$$

$$\frac{d^2F}{dt^2}(0.524) = -\sin 0.524 = -0.500$$

$$\frac{d^3F}{dt^3}(0.524) = -\cos 0.524 = -0.866 \cdot$$

Therefore,

$$\sin t = 0.500 + 0.866(t - 0.524) - \frac{0.500}{2!}(t - 0.524)^2 - \frac{0.866}{3!}(t - 0.524)^3 + \ldots$$

Calculating for t = 32° = 0.559 rad,

The first term is + 0.500.

The second term is + 0.866 x (0.559 − 0.524) = 0.030.

The third term is $- \dfrac{0.500}{2}(0.559 - 0.524)^2 = -0.00031$.

There is no merit to calculating additional terms since they are not large enough to have any bearing on the value of sin 32° to three decimal places.

Accordingly, sin 32° = 0.500 + 0.030 = 0.530.

5

Complex Quantities

Background

A quadratic equation involving the variable "x" can be written in its general form and can then be solved using an algebraic procedure. The general form of the quadratic equation in x, is

$$a x^2 + b x + c = 0.$$

Because of the quadratic (second power) nature of the equation, two values of x will satisfy it. These values of x are called, in mathematical parlance, the "roots of the equation." They can be designated m_1 and m_2, and their values are

$$m_1 = \frac{-b + \sqrt{b^2 - 4ac}}{2a} \text{ and } m_2 = \frac{-b - \sqrt{b^2 - 4ac}}{2a}$$

The solution for any quadratic equation can consequently be found by applying this formula for m_1 and m_2. For example, given that

$$x^2 + x - 6 = 0 \text{ (i.e., } a = 1, b = 1, c = -6),$$

$$\text{then } m_1 = \frac{-1 + \sqrt{1 + 24}}{2} = \frac{-1 + 5}{2} = 2,$$

$$\text{and } m_2 = \frac{-1 - \sqrt{1 + 24}}{2} = \frac{-1 - 5}{2} = -3.$$

Therefore, $x = 2$ and $x = -3$ are the solutions for $x^2 + x - 6 = 0$.

The determination of the roots of a quadratic equation is straightforward, provided that the quantity under the square root sign ($b^2 - 4ac$) does not turn out to be negative, as it would if the equation to be solved were

$$x^2 - 6x + 13 = 0.$$

In this case,

$$m_1 = \frac{+6 + \sqrt{36 - 52}}{2} = \frac{6 + \sqrt{-16}}{2} \text{ and } m_2 = \frac{6 - \sqrt{-16}}{2}.$$

The term $\sqrt{-16}$ can be simplified one step further by taking the factor 4 outside of the square root sign, leaving only the factor (-1). That is, $\sqrt{-16} = 4\sqrt{-1}$. The roots of the equation then become

$$m_1 = \frac{6 + 4\sqrt{-1}}{2} = 3 + 2\sqrt{-1}, \text{ and } m_2 = \frac{6 - 4\sqrt{-1}}{2} = 3 - 2\sqrt{-1}$$

This introduces the concept of a "number" whose value is $\sqrt{-1}$. This number will be identified by the letter j, that is, $j = \sqrt{-1}$. Since $\sqrt{-1}$ cannot be evaluated, or from another viewpoint, it is not possible to draw a line that is $\sqrt{-1}$ units long, the quantity j must be imaginary.

It is important to understand, however, that as far as mathematics is concerned, stating that a number is *imaginary* is entirely different from stating that there is *no such number*. The quantity j is imaginary because it cannot be observed in nature, but it definitely exists because $x = (3 + 2j)$ and $x = (3 - 2j)$ are values that satisfy the equation $(x^2 - 6x + 13) = 0$. This can be proven by substituting the value $x = (3 + 2j)$ into the original equation, bearing in mind that $j^2 = -1$.

$$x^2 = 9 + 12j + 4j^2 = 9 + 12j - 4 = 5 + 12j$$

$$-6x = -18 - 12j$$

$$+13 = 13$$

$$x^2 - 6x + 13 = 5 + 12j - 18 - 12j + 13 = 0$$

The numbers $(3 + 2j)$ and $(3 - 2j)$, which in this case are the roots of $(x^2 - 6x + 13) = 0$, are called *complex numbers*. The characteristic of a complex number is that it contains an imaginary part, that is, a part that contains the number j. A complex number usually contains a real part as well, as it does in the case of the complex number $(3 + 2j)$. In this instance, the real part is 3 and the imaginary part is 2j.

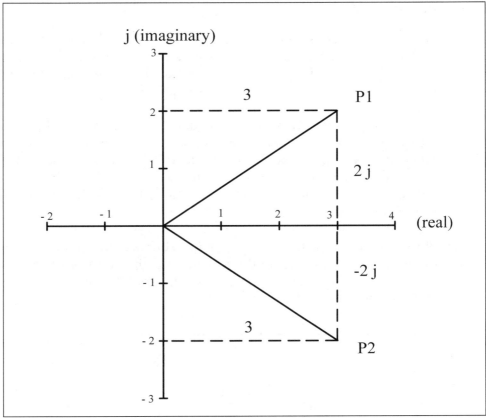

Figure 5-1. Graphical Representation of Complex Numbers.

The real part of a complex number may be zero, leaving only the imaginary part. If the imaginary part were zero, however, only the real part would remain, and the number would be a real number rather than a complex number.

Graphical Representation

A complex number can be represented graphically by plotting it on a complex plane, in which the axis for the real part is horizontal, while the axis for the imaginary part is vertical. In Figure 5-1, the point P1 represents the complex number (3 ɪ 2 j), while P2 represents (3 – 2 j).

When the roots of a quadratic equation with real coefficients are complex numbers, they will always occur in pairs, called *conjugate pairs*. This means that the roots are of the form (a + jb) and (a – jb). In any conjugate pair, the real parts of both numbers are the same, while the imaginary parts differ in sign only.

The Complex Variable

Variables, as well as numbers, can be complex, which means they can have an imaginary part. If z is a complex variable, then it *may* have a real part, but it *will* have an imaginary part. The real and imaginary parts can be designated by x and y, respectively, so that z = (x + j y).

Figure 5-2 is the graphical representation of a complex variable in both rectilinear and polar coordinates. In rectilinear coordinates, the x component is shown measured along the horizontal (real) axis, while the y component is measured along the vertical (imaginary or j) axis. The sum of x horizontally and jy vertically, which is z, creates a vector, which starts at the origin O and ends at P.

Since z has not only the quality of length but also the quality of direction, depending on the values of the real and imaginary parts x and y, the vector z is designated $\overline{OP}$. The bar over the letters is the shorthand symbol that indicates that OP is a vector.

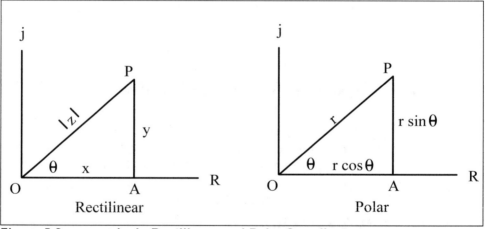

Figure 5-2. z = x + j y in Rectilinear and Polar Coordinates.

The length, or magnitude, of the vector OP is equal to the distance OP and is customarily identified by $|z|$. In mathematics terminology, $|z|$ is called the *modulus* of the complex number z. From the triangle OAP, $|z|^2 = x^2 + y^2$. Therefore,

$$|z| = \sqrt{x^2 + y^2}.$$

The vector OP is vector the sum of x + j y in a graphical representation using rectilinear coordinates. Certain problems may be dealt with more conveniently by representing a complex number in a system of polar coordinates. In the polar system, the modulus of z is equal to r. The direction of the vector OP is shown by the angle θ, which is the angle formed by OP and the horizontal or real axis. This angle is defined by the relation

$$\tan\theta = \frac{y}{x}, \text{ or } \theta = \tan^{-1}\frac{y}{x} \left(\text{the angle whose tangent is } \frac{y}{x}\right).$$

The angle θ is called the *argument* of the vector z, or *arg z* for short.

These two diagrams show that $x = r\cos\theta$, and $y = r\sin\theta$.

Trigonometric and Exponential Functions

If the MacLaurin series expansion is used to develop an infinite series expression for $e^{j\theta}$ (bearing in mind that $j^2 = -1$), the result is

$$e^{j\theta} = 1 + j\theta - \frac{\theta^2}{2!} - j\frac{\theta^3}{3!} + \frac{\theta^4}{4!} + j\frac{\theta^5}{5!} - \frac{\theta^6}{6!} - j\frac{\theta^7}{7!} + \dots$$

$$= \left(1 - \frac{\theta^2}{2!} + \frac{\theta^4}{4!} - \frac{\theta^6}{6!} + \dots\right) + j\left(\theta - \frac{\theta^3}{3!} + \frac{\theta^5}{5!} - \frac{\theta^7}{7!} + \dots\right).$$

The expression in first set of brackets is the MacLaurin series for $\cos\theta$, while the expression in the second set of brackets is the series for $\sin\theta$. Consequently,

$$e^{j\theta} = \cos\theta + j\sin\theta.$$

By using the same technique, it can be shown that $e^{-j\theta} = \cos\theta - j\sin\theta$.

If these expressions are added, $e^{j\theta} + e^{-j\theta} = 2\cos\theta$, and

$$\cos\theta = \frac{e^{j\theta} + e^{-j\theta}}{2}.$$

If the second expression is subtracted from the first,

$$e^{j\theta} - e^{-j\theta} = 2j\sin\theta, \text{ and } \sin\theta = \frac{e^{j\theta} - e^{-j\theta}}{2j}.$$

These relationships frequently come in handy in the solution of differential equations.

Sum of Two Complex Quantities

If $z_1 = x_1 + j\, y_1$, and $z_2 = x_2 + j\, y_2$, then

$$z_1 + z_2 = x_1 + j\, y_1 + x_2 + j\, y_2 = (x_1 + x_2) + j\, (y_1 + y_2).$$

This indicates that to find the sum of two complex quantities, the real and imaginary parts should be added separately. Figure 5-3 illustrates this.

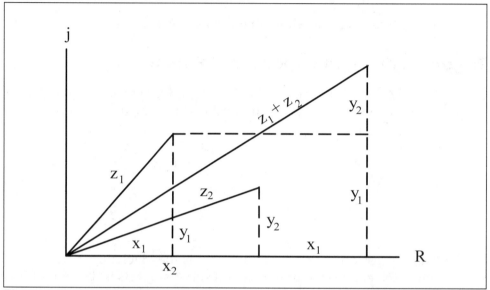

Figure 5-3. Graphical Representation of the Sum of Two Complex Quantities.

Product of Two Complex Quantities

Figure 5-4 is a graphical illustration of the product of two complex quantities.

The product of two complex quantities $z_1 = x_1 + j\, y_1$ and $z_2 = x_2 + j\, y_2$ is most easily determined by converting to the polar form.

If $z_1 = r_1 e^{j\theta_1}$ and $z_2 = r_2 e^{j\theta_2}$, then

$$
\begin{aligned}
z_1 \times z_2 &= r_1 e^{j\theta_1} \times r_2 e^{j\theta_2} \\
&= r_1 r_2 e^{j(\theta_1 + \theta_2)} \\
&= |z_1||z_2| e^{j(\theta_1 + \theta_2)}
\end{aligned}
$$

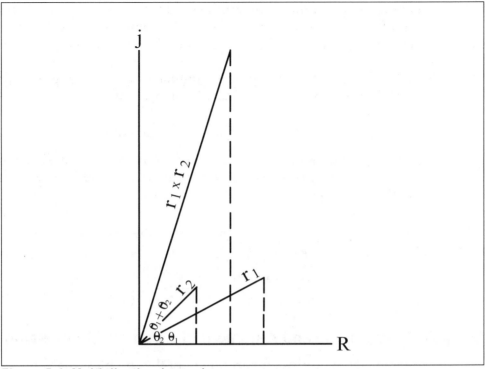

Figure 5-4. Multiplication (z_1 x z_2).

This verifies that the product of two complex quantities is obtained by multiplying the magnitudes and adding the arguments.

Separating the Real and Imaginary Parts

It is often necessary to rearrange the terms of a complex quantity to determine the magnitude and the argument. This is done by collecting the terms that are *not* multiplied by j into one group, and the terms that *are* multiplied by j into another. For example, suppose $z = (a + jb) (c + jd)$. Multiplying this yields

$$z = ac + jad + jbc - bd.$$

The real part is $(ac - bd)$, while the imaginary part is $(ad + bc)$.

The modulus $|z| = \sqrt{(ac - bd)^2 + (ad + bc)^2}$.

The argument $\theta = \tan^{-1}\left(\dfrac{ad + bc}{ac - bd}\right)$.

Separation of the complex quantity into its real and imaginary parts is straightforward, except in cases where the complex quantity is a fraction. When this happens, the algebraic relation $(a - b)(a + b) = (a^2 - b^2)$ is used.

Notice that if b were an imaginary number, ($= jc$), the imaginary part would disappear when the product is taken. That is,

$$(a + jc)(a - jc) = (a^2 - j^2 c^2) = (a^2 + c^2).$$

In this way the denominator of the fraction can be cleared of imaginary numbers, and the fraction can be separated in real and imaginary parts using the procedure

$$\frac{x + jy}{a + jb} = \frac{x + jy}{a + jb} \times \frac{a - jb}{a - jb} = \frac{xa - jxb + jya + yb}{a^2 + b^2}$$

$$= \frac{xa + yb + j(ya - xb)}{a^2 + b^2} = \frac{xa + yb}{a^2 + b^2} + j\left(\frac{ya - xb}{a^2 + b^2}\right).$$

Example 1: Magnitude and Argument of a Complex Expression

Determine the modulus and argument of the complex quantity

$$z = \frac{A}{j\omega t + 1}.$$

To clear the denominator of the imaginary number, multiply the numerator and denominator by $(j\omega t - 1)$.

$$z = \frac{A}{j\omega t + 1} \times \frac{j\omega t - 1}{j\omega t - 1} = \frac{Aj\omega t - A}{-\omega^2 t^2 - 1} = \frac{A - Aj\omega t}{\omega^2 t^2 + 1}$$

$$= \frac{A}{\omega^2 t^2 + 1} - j\frac{A\omega t}{\omega^2 t^2 + 1}.$$

The modulus $|z| = \sqrt{\dfrac{A^2}{(\omega^2 t^2 + 1)^2} + \dfrac{A^2 \omega^2 t^2}{(\omega^2 t^2 + 1)^2}}$

$$= \frac{A}{\omega^2 t^2 + 1}\sqrt{1 + \omega^2 t^2} = \frac{A}{\sqrt{1 + \omega^2 t^2}}.$$

The argument $\theta = \tan^{-1}\left(-\dfrac{A\omega t}{\omega^2 t^2 + 1} \div \dfrac{A}{\omega^2 t^2 + 1}\right)$

$\qquad\qquad = \tan^{-1}\left(-\dfrac{A\omega t}{A}\right) = \tan^{-1}(-\omega t).$

6

Differential Equations

Introduction

There is an area of mathematics that deals with equations that contain derivatives of a variable with respect to another variable, or variables. These equations are called differential equations. The following are examples.

$$(1) \qquad \frac{1}{k}\frac{dx}{dt} + x = X$$

$$(2) \qquad m\frac{d^2x}{dt^2} + r\frac{dx}{dt} + kx = Q$$

$$(3) \qquad 1 + \left(\frac{dx}{dt}\right)^2 = 3\frac{d^3x}{dt^3}$$

$$(4) \qquad 1 + \left(\frac{d^2x}{dt^2}\right)^2 = 3x + \frac{dx}{dt}$$

$$(5) \qquad x^2\frac{dx}{dt} + tx = 10$$

$$(6) \qquad \frac{\partial^2 x}{\partial t^2} = a^2\frac{\partial^2 y}{\partial x^2}$$

The solution of a differential equation requires that an equation be obtained that has the variables in their natural form, that is, free of all derivatives. In the case of the first example above, the solution is:

$$x = X\left(1 - e^{-kt}\right)$$

This solution can be verified. If it is actually the original relation between x and t, then taking the derivative with respect to t of both sides gives

$$\frac{dx}{dt} = \frac{d}{dt}(X - Xe^{-kt}) = Xke^{-kt}.$$

Then the left side of the differential equation equals

$$\frac{1}{k}\frac{dx}{dt} + x = \frac{1}{k}Xke^{-kt} + X(1 - e^{-kt}) = Xe^{-kt} + X - Xe^{-kt} = X$$

which is the right side of the equation.

Philosophy

In many cases, a set procedure cannot be established for the solution of a particular differential equation. In fact, many differential equations, principally those that lack a certain degree of symmetry or orderliness, are incapable of solution. Solving differential equations is often as much an art as it is a science. Mathematical intuition and experience are valuable assets.

Solvable differential equations tend to fall into patterns, so that part of the skill required to solve a differential equation lies in being able to spot the pattern and in knowing the right procedure for dealing with it.

As an example, the motion of a mass suspended from a spring and caused to bounce up and down can be described by the differential equation in Example (2). Observing the motion of the mass reveals, first of all, that the solution must contain a cyclic factor such as $\sin \omega t$, which describes the up and down movement, and another factor e^{-at}, which describes the gradual dying out of the oscillations over a period of time. In dealing with this differential equation, it is helpful to know at the outset that the solution must look something like $y = Ce^{-at}\sin \omega t$.

Definitions

An *ordinary* differential equation is one with just one independent variable. The independent variable is nearly always the variable that appears in the denominator of the derivative term. Only total derivatives are present; there are no partial derivatives.

Examples 1 through 5 are ordinary differential equations, while Example 6 is a partial differential equation with two independent variables.

The *order* of a differential equation is the order (the number of times the derivative has been taken) of the highest order derivative in the equation.

Examples 1 and 5 are consequently differential equations of the first order. Examples 2, 4, and 6 are of the second order. Example 3 is of the third order.

The *degree* of the differential equation is the degree (power) to which the highest order derivative in the equation has been raised. Note that the degree of the equation is not necessarily established by the highest power term that appears in the equation.

Examples 1, 2, 5, and 6 are differential equations of the first degree. Example 3 is also a first degree equation because, although one of the derivative terms is raised to the second power, this derivative is not the highest order derivative in the equation. The highest order derivative,

$$\frac{d^3t}{dt^3}$$

is raised to the first power only. Example 4, however, is a second degree equation.

A differential equation is *linear* if the coefficients multiplying the various derivative terms are constants, or at worst, functions of the independent variable. In addition, the power of each of the derivative terms can be no higher than one. This means that a linear differential equation must be of the first degree.

Examples 1, 2, and 6 are all linear differential equations. Examples 3 and 4 are nonlinear because of the derivative terms which are squared. Example 5 is nonlinear because of the factor x^2 (a function of the dependent variable), which multiplies the first term.

A linear differential equation has a general form, which is

$$f_n(t)\frac{d^n x}{dt^n} + f_{n-1}(t)\frac{d^{n-1}x}{dt^{n-1}} + f_{n-2}(t)\frac{d^{n-2}x}{dt^{n-2}} + \ldots$$

$$\ldots + f_2(t)\frac{d^2 x}{dt^2} + f_1(t)\frac{dx}{dt} + f_0(t)x = f(t).$$

Application

Control systems engineering pertains to the study of the dynamics of the components of control systems, including both the process being controlled and the items of control hardware, for the purpose of obtaining satisfactory overall dynamics when these components are combined into a control system. A control system that has good dynamic behavior will recover quickly from upsets and will generally result in close automatic control.

Dynamics is a descriptive term, which characterizes the reaction of control system components, or complete systems, to impulses that vary with time. The dynamic behavior of many control system components and systems can be described by differential equations in which the independent variable is time. A knowledge of how to resolve these differential equations is consequently valuable in control systems engineering analysis.

Fortunately, the less complicated types of differential equations are frequently the ones that are involved with control system analysis. The most commonly encountered equation is likely the linear differential equation with constant coefficients. In its most general form, this equation would appear as

$$\frac{d^n x}{dt^n} + a_{n-1}\frac{d^{n-1}x}{dt^{n-1}} + a_{n-2}\frac{d^{n-2}x}{dt^{n-2}} + ... + a_2\frac{d^2 x}{dt^2} + a_1\frac{dx}{dt} + a_0 x = f(t).$$

In this equation the a's are constants.

Differential Equations of the First Order and First Degree

The general form of this type of differential equation is

$$P\frac{dx}{dt} + Q = 0$$

where P and Q are both functions of t and x.

Consequently, this type of equation, in its general form, is not necessarily linear.

It is not possible to solve the general form of this equation and arrive at a formula that will give an automatic answer for any particular problem. When the problem is specifically known, however, the solution (assuming this is possible) will usually be obtainable because the equation belongs in one of four categories.

Category 1: Exact Differentials

This means that the expression

$$P\frac{dx}{dt} + Q$$

is actually the derivative of some other expression R, where R is a function of t, x, or both. In other words,

$$P\frac{dx}{dt} + Q = \frac{dR}{dt} = 0.$$

The solution is R = C, where C is a constant.

This method of arriving at the solution to the differential equation is not particularly useful except for full time mathematicians who have the experience to spot the exact differential. Furthermore, it turns out that equations that are exact differentials can also be solved by other means, whether or not it is recognized that they actually are exact differentials.

Category 2: Variables Separable

It may be possible, through a rearrangement of the terms, to get all of the x (dependent variable) terms on the left side of the equals sign and all of the t (independent variable) terms on the right side. The variables are then separated, and the problem is reduced to integrating the expressions on either side of the equation.

Accordingly, a logical starting point in the solution of a first degree, first order, differential equation, is to determine if it is possible to separate the variables.

The first example in the set of examples on page 73 describes the output behavior of a control system component whose reaction, designated by x, goes from x = 0 to x = X as a result of an impulse which has been applied to it. The output change from 0 to X is not instantaneous, however. The rate of change of x at any time t is proportional to the difference between the value of x at that instant and its ultimate value X. Since as time goes on the difference (X − x) is diminishing, the rate of change of x will fall off accordingly. Components with this type of dynamic behavior are quite common in control systems and are generally referred to as *time constants*.

This differential equation can be rearranged as $\frac{dx}{dt} = k(X - x)$.

In this equation the variables can be separated.

$$\frac{dx}{X-x} = k\,dt, \text{ so that } \int\frac{1}{X-x}dx = -\int\frac{1}{X-x}d(X-x) = k\int dt.$$

Integrating both sides yields,

$$-\log_e(X-x) = kt + \log_e C$$

The constant of integration is required here. Since C is a constant, then $\log_e$ C will also be a constant. Inasmuch as there is already a logarithm in the equation, the log form of the constant will make it easier to manipulate the equation into its final form.

$$\log_e C + \log_e (X-x) = -kt = \log_e \{C \times (X-x)\}$$

$$C(X-x) = e^{-kt} \text{ and } x = X - \frac{1}{C}e^{-kt}.$$

If $x = 0$ when $t = 0$, then $0 = X - \frac{1}{C}$ and $\frac{1}{C} = X$. Therefore,

$$x = X - Xe^{-kt} = X(1 - e^{-kt}).$$

Category 3: Homogeneous Equations

When a differential equation has the form

$$\frac{dx}{dt} = f\left(\frac{x}{t}\right),$$

it is termed a *homogeneous* differential equation, for reasons that are presumably obvious to qualified mathematicians, but not to the average math student.

The test for a homogeneous equation is to substitute the product vt for x in the right side of the equation. Since vt = x, then the new variable v = x/t. If the right side expression is actually a function of x/t, then after substituting vt = x in this expression, the t's will cancel out, leaving an expression containing v only.

For example, given the differential equation $2t^2\frac{dx}{dt} - x^2 = t^2$, which in rearranged form is

$$\frac{dx}{dt} = \frac{t^2 + x^2}{2t^2}.$$

In this equation, the variables cannot be separated. However, replacing x with vt in the expression on the right side gives

$$\frac{dx}{dt} = \frac{t^2 + v^2 t^2}{2t^2} = \frac{1 + v^2}{2}.$$

Since the t terms cancel out completely, leaving only the v terms, the equation is homogeneous.

The substitution vt = x, which is used as the test for homogeneity, is also worth trying as a solution for the differential equation. If x is made equal to vt, then,

$$\frac{dx}{dt} = \frac{d}{dt}(vt) = v + t\frac{dv}{dt}.$$

Substituting vt for x in both sides of the equation accordingly gives

$$v + t\frac{dv}{dt} = \frac{1 + v^2}{2},$$

in which the variables can be separated.

$$t\frac{dv}{dt} = \frac{1 + v^2}{2} - v = \frac{1 + v^2 - 2v}{2} = \frac{(1-v)^2}{2}, \text{ so that}$$

$$\int \frac{dt}{t} = 2\int \frac{1}{(1-v)^2}dv = -2\int \frac{1}{(1-v)^2}d(1-v) = -2\int (1-v)^{-2}d(1-v).$$

Performing the integration, with C as the constant of integration,

$$\log_e t = -2\left(\frac{-1}{1-v}\right) + \log_e C, \text{ so } \log_e \frac{t}{C} = \frac{2}{1-v}$$

$$1 - v = \frac{2}{\log_e \frac{t}{C}} \text{ and } v = 1 - \frac{2}{\log_e \frac{t}{C}} = \frac{x}{t}$$

$$\text{Therefore, } x = t\left(1 - \frac{2}{\log_e \frac{t}{C}}\right).$$

Category 4: Linear Differential Equations

A differential equation of the first order and first degree, which is linear in addition, would have the general form

$$\frac{dx}{dt} + Px = Q$$

where P and Q are constants or functions of t but *not* of x. In this case, it is possible to obtain a general solution by the following method.

It was shown previously that a differential equation to be solved will sometimes turn out to be an exact differential. To obtain a solution for the general equation

$$\frac{dx}{dt} + Px = Q,$$

the approach is to find a factor R, which is a function of t (only), such that when each of the terms of the equation is multiplied by R, the left side of the equation becomes an exact differential. Accordingly,

$$R\frac{dx}{dt} + RPx = RQ, \text{ and } R\frac{dx}{dt} + RPx$$

is to be an exact differential. Notice the similarity between

$$\left\{ R\frac{dx}{dt} + RPx \right\} \text{ and } \left\{ R\frac{dx}{dt} + x\frac{dR}{dt} \right\}, \text{ which happens to be } \frac{d}{dt}(Rx).$$

This suggests that the exact differential required is $\frac{d}{dt}(Rx)$, provided that $\frac{dR}{dt}$ is equal to RP.

If $\frac{dR}{dt} = RP$, then $\int \frac{dR}{R} = \int P\,dt$, and $\log_e R = \int P\,dt$.

$$\text{Therefore, } R = e^{\int P\,dt}.$$

The complete solution is $\frac{d}{dt}(Rx) = RQ,$

$$\int d(Rx) = \int RQdt$$

$$Rx = \int RQdt$$

$$\text{and } x = \frac{1}{R}\int RQdt, \text{ where } R = e^{\int Pdt}.$$

Example 1: Time Constant

The differential equation that has already been examined is that in which the dependent variable y, as a result of a step change impulse, is changing from its starting point y = 0 so as to eventually attain a new value Y. However, the rate of change of y is proportional to (Y − y), so that it is constantly diminishing as y approaches Y. The differential equation is

$$\frac{dy}{dt} = k(Y - y)$$

where k is the proportional constant. Rearranging this,

$$\frac{dy}{dt} + ky = kY.$$

Thus, in this relation, P = k, and Q = kY.

$$\int P\,dt = \int k\,dt = kt, \text{ so } R = e^{kt}.$$

Therefore, applying the formula,

$$y = \frac{1}{e^{kt}}\int e^{kt}kY dt = \frac{kY}{e^{kt}}\int e^{kt}dt = \frac{kY}{e^{kt}}\left(\frac{1}{k}e^{kt} + C\right) = Y\left(1 + \frac{kC}{e^{kt}}\right)$$

where C is the constant of integration.

Since it is known that y = 0 when t = 0, the value of C can be found by substituting these values in the equation for y.

$$0 = Y\left(1 + \frac{kC}{1}\right)$$

from which $C = -\frac{1}{k}$.

Therefore, the final solution is

$$y = Y\left[1 + k\left(\frac{-1}{k}\right)\left(\frac{1}{e^{kt}}\right)\right] \text{ or } y = Y(1 - e^{-kt}).$$

Linear Differential Equations with Constant Coefficients

A first order equation of this type would have the form

$$\frac{dx}{dt} + a_0 x = f(t) \ \ (a_0 \text{ is a constant}).$$

The situation in which $f(t) = 0$ will be considered first.

The equation $\frac{dx}{dt} + a_0 x = 0$ can be solved by separating the variables.

$$\frac{dx}{dt} = -a_0 x, \text{ thus } \frac{dx}{x} = -a_0 dt, \text{ and } \int \frac{dx}{x} = -a_0 \int dt$$

$$\log_e x = -a_0 t + \log_e C \ \ (\text{where C is the constant of integration})$$

$$\log_e \frac{x}{C} = -a_0 t$$

$$\frac{x}{C} = e^{-a_0 t} \text{ and } x = C e^{-a_0 t}$$

Second Order Linear Differential Equation with Constant Coefficients

A second order linear differential equation with constant coefficients would have the form

$$a_2 \frac{d_2 x}{dt^2} + a_1 \frac{dx}{dt} + a_0 x = f(t).$$

This differential equation actually describes systems in the real world that can oscillate. In closed loop control systems, part of the output of the process is fed back to the input of the system as a measurement signal. This sets up the conditions that are conducive to oscillation.

In control systems studies, however, it is generally considered that oscillations in the system are triggered by a single transient input at time zero. Everything that happens from then on depends on the nature of the system itself, not on any further external influences. The net result is that in the differential equation, f(t) can be considered to be zero.

Furthermore, if f(t) = 0, then all of the terms on the left side can be divided by the constant a_2, which reduces the number of constants from 3 to 2. The two new constants will be A (replacing a_1/a_2) and B, (replacing a_0/a_2).

A solution will consequently be sought for

$$\frac{d^2x}{dt^2} + A\frac{dx}{dt} + Bx = 0.$$

At this point the procedure becomes less orderly and more abstract, in that the will of the wisp of mathematical intuition gets involved. In fact, it frequently turns out that the solution to a particular differential equation is found because someone with the right mathematical background is able to guess at the answer, and then verify that he or she was right. In this case, it was already shown that the solution for the first order equation with f(t) = 0 was

$$x = Ce^{-a_0 t}.$$

This suggests that $x = Ce^{mt}$, with the value of m to be determined, may be a solution for the second order equation as well.

If this trial solution is valid, then

$$x = Ce^{mt}, \quad \frac{dx}{dt} = mCe^{mt}, \quad \text{and} \quad \frac{d^2x}{dt^2} = m^2Ce^{mt}.$$

Inserting these values in the original equation yields

$$m^2Ce^{mt} + AmCe^{mt} + BCe^{mt} = 0, \text{ which reduces to}$$

$$m^2 + Am + B = 0.$$

This means that the trial solution could be a solution, provided that it is possible to find a value for m that satisfies the algebraic equation $m^2 + Am + B = 0$. As it turns out, there are actually two values of m (to be designated m_1 and m_2), which will satisfy this requirement. These are

$$m_1 = \frac{-A + \sqrt{A^2 - 4B}}{2}, \text{ and } m_2 = \frac{-A - \sqrt{A^2 - 4B}}{2}.$$

A complete solution for

$$\frac{d^2x}{dt^2} + A\frac{dx}{dt} + Bx = 0$$

will then be $x = C_1e^{m_1t} + C_2e^{m_2t}$.

Realistically, this should be verified.

$$x = C_1e^{m_1t} + C_2e^{m_2t}, \quad \frac{dx}{dt} = m_1C_1e^{m_1t} + m_2C_2e^{m_2t}, \text{ and}$$

$$\frac{d^2x}{dt^2} = m_1^2C_1e^{m_1t} + m_2^2C_2e^{m_2t}$$

Inserting these expressions in the original equation results in

$$\left\{ m_1^2\,C_1\,e^{m_1t} + m_2^2\,C_2\,e^{m_2t} \right\} + A\left\{ m_1\,C_1\,e^{m_1t} + m_2\,C_2\,e^{m_2t} \right\}$$

$$+ B\left\{ C_1\,e^{m_1t} + C_2\,e^{m_2t} \right\},$$

which is equal to

$$C_1\,e^{m_1t}\left(m_1^2 + A_1\,m_1 + B \right) + C_2\,e^{m_2t}\left(m_2^2 + A\,m_2 + B \right)$$

$$= \left(C_1\,e^{m_1t} \times 0 \right) + \left(C_2\,e^{m_2t} \times 0 \right)$$

$$= \text{zero} = \text{right side of the equation.}$$

Inasmuch as the solution of a second order differential equation will involve two integrations, the general solution should contain two arbitrary constants.

The solution $x = C_1e^{m_1t} + C_2e^{m_2t}$ satisfies this requirement. It can therefore be considered to be the general solution.

The equation $m^2 + A\,m + B = 0$, from which the values of m_1 and m_2 are determined, is called the *auxiliary equation*. The dynamic behavior that is described by the differential equation depends on what the values of m_1 and m_2 turn out to be. As has been shown, these values are

$$m_1 = \frac{-A + \sqrt{A^2 - 4B}}{2} \quad \text{and} \quad m_2 = \frac{-A - \sqrt{A^2 - 4B}}{2}.$$

If A^2 is greater than 4B, then m_1 and m_2 will be real numbers, although either, or both, could be negative. In any event, the solution would be the sum of two expressions in t, which are changing exponentially with time. The absence of a sine or cosine function is an indication that the system does not oscillate.

There is also the possibility that $A^2 = 4B$, in which case m_1 and m_2 are equal. The solution then is

$$x = (C_1 + C_2)e^{mt}$$

where m is the common value of m_1 and m_2. Here again the system behaves in exponential fashion and does not oscillate. Consequently, neither of these solutions is of great interest to students of automatic control systems.

The Oscillatory Case

The final possibility is the situation in which A^2 is less than 4B. In this case, the square root of a negative number is involved, and both roots of the auxiliary equation are complex numbers, namely,

$$m_1 = \frac{-A + j\sqrt{4B - A^2}}{2} \text{ and } m_2 = \frac{-A - j\sqrt{4B - A^2}}{2}$$

with $j = \sqrt{-1}$.

In addition to being complex, the roots m_1 and m_2 occur in conjugate pairs; that is, they are of the form $m_1 = \alpha + j\omega$ and $m_2 = \alpha - j\omega$, where

$$\alpha = \frac{-A}{2}, \text{ and } \omega = \frac{\sqrt{4B - A^2}}{2}.$$

It will be easier to work with $m_1 = \alpha + j\omega$ and $m_2 = \alpha - j\omega$ until the final answer is reached, and then replace α and ω with the original factors A and B.

Proceeding with the solution:

$$x = C_1 e^{m_1 t} + C_2 e^{m_2 t} = C_1 e^{(\alpha + j\omega)t} + C_2 e^{(\alpha - j\omega)t}$$

$$= C_1 e^{\alpha t} e^{j\omega t} + C_2 e^{\alpha t} e^{-j\omega t} = e^{\alpha t}\left(C_1 e^{j\omega t} + C_2 e^{-j\omega t}\right).$$

In general, $e^{j\omega t} = \cos \omega t + j \sin \omega t$, and $e^{-j\omega t} = \cos \omega t - j \sin \omega t$. Using these relationships,

$$x = e^{\alpha t}\left[C_1\left(\cos\omega t + j\sin\omega t\right) + C_2\left(\cos\omega t - j\sin\omega t\right)\right]$$

$$= e^{\alpha t}\left[\left(C_1 + C_2\right)\cos\omega t + j\left(C_1 - C_2\right)\sin\omega t\right].$$

The Constant of Integration

The solution for a differential equation sooner or later requires the integration of some expression. For the integration to produce a general result, the result has to include a constant of integration. This is because mathematical expressions that differ only by a constant will all have the same derivative. The appropriate value for the constant (or constants, as in this case) is usually determined in the final step by applying initial conditions.

The fact that the integral must include the constant does not mean, however, that the constant has to be an ordinary number. Any form of the constant is valid, provided that the form itself is basically a constant. This simply means that if C is a constant, then so are $(-C)$, C^2, the square root of C, e^C, $\log_e C$, $\sin C$, and so on. It is also possible for C to be a complex number.

One edge that accomplished mathematicians have is the ability to visualize the format that the constant should take, so that the final solution of the differential equation will be in its most useful form.

In the problem at hand, it would be desirable to have the expression inside of the box brackets in the form $(\sin\phi\cos\omega t + \cos\phi\sin\omega t)$, since this is equal to $\sin(\omega t + \phi)$. This would be possible if the constants C_1 and C_2 were replaced in the solution by two new parameters X and ϕ, such that $C_1 + C_2 = X\sin\phi$, and $j(C_1 - C_2) = X\cos\phi$.

What now needs to be shown is that given the expressions $(C_1 + C_2)$ and $j(C_1 - C_2)$ above, C_1 and C_2 each has its own value, separate from the other. This can be verified if $\sin\phi$ and $\cos\phi$ are converted to their exponential form.

$$X\sin\phi = X\left(\frac{e^{j\phi} - e^{-j\phi}}{2j}\right) = C_1 + C_2$$

$$X\cos\phi = X\left(\frac{e^{j\phi} + e^{-j\phi}}{2}\right) = j\left(C_1 - C_2\right)$$

From (1) $Xe^{j\phi} - Xe^{-j\phi} = 2jC_1 + 2jC_2$

From (2) $Xe^{j\phi} + Xe^{-j\phi} = 2jC_1 - 2jC_2$

Adding these expressions: $2Xe^{j\phi} = 4jC_1$, and $C_1 = \dfrac{Xe^{j\phi}}{2j}$.

Subtracting, $-2Xe^{-j\phi} = 4jC_2$, and $C_2 = -\dfrac{Xe^{-j\phi}}{2j}$.

Once it has been verified that C_1 and C_2 have their own values, even though the values may be complex numbers, then the solution for the original differential equation becomes

$$x = e^{\alpha t}\left[(C_1 + C_2)\cos\omega t + j(C_1 - C_2)\sin\omega t\right]$$

$$= Xe^{\alpha t}\left(\sin\phi\,\cos\omega t + \cos\phi\,\sin\omega t\right)$$

$$= Xe^{\alpha t}\sin(\omega t + \phi).$$

Commentary on the Result

The expression $x = Xe^{\alpha t}(\sin\omega t)$ is the oscillatory solution for the differential equation

$$\frac{d^2x}{dt^2} + A\frac{dx}{dt} + Bx = 0.$$

In the real world, x may represent a temperature, pressure, or voltage displacement of an object, or some other variable whose value must be tracked. The sine function in the result says that the value of x will oscillate up and down. The fact that the right side of the original differential equation is zero says that after the original disturbance that starts the oscillations takes place, the system is allowed to oscillate on its own. It is not driven or further disturbed by any external force or influence.

The values of the parameters α and ω are established by the constants A and B in the original equation. Their values are

$$\alpha = -\frac{1}{2}A, \text{ and } \omega = \frac{1}{2}\sqrt{4B - A^2}.$$

ω is the frequency at which the system will oscillate. If time is measured in seconds, the units of ω will be radians per second.

α is the modifier that determines whether the oscillations get bigger or smaller. If α is >1, the oscillations increase. If α is <1, but not zero, the oscillations decrease and eventually die out. This is the situation that is sought in control systems.

If $\alpha = 0$, then the $e^{\alpha t}$ term becomes 1 and the oscillations go on forever. Since α depends only on the constant A, this also implies that A is zero, and the original differential equation has the form

$$\frac{d^2x}{dt^2} + Bx = 0.$$

The X and ϕ terms will be determined by the initial conditions, that is, the conditions that exist at $t = 0$. X is the amplitude of the first oscillation, to be modified subsequently by the value of α.

Finally, ϕ is the phase displacement of the oscillations on the time scale. The value of ϕ establishes the point where the oscillations will be, at maximum, minimum, zero, or whatever, when t is zero.

Example 2: The Spring/Mass System

Figure 6-1 portrays a spring that is fastened at its top end, and with a mass attached to its lower end. Under this arrangement the mass is free to oscillate up and down. However, attached to the mass is a dashpot, which imposes some resistance to the movement of the mass, with the result that the oscillations will eventually die out. The following data are known.

- The mass of the moving mass is m kg.
- The constant of the spring is k Newtons per metre.
- The resistance coefficient of the dashpot is r Newtons per metre per second.
- The elevation of the mass at any instant is y metres with respect to the $y = 0$ base line.

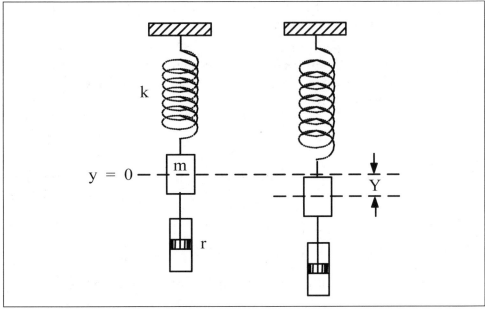

Figure 6-1. The Spring/Mass System.

The requirement is to find an expression of the form y = f(t), which will pinpoint the position of the mass at any time t.

At the outset, it will be specified that elevations above the base line and upward forces are positive, while elevations below the base line and downward forces are negative.

When the mass is at rest, y = 0, there are only two forces affecting it. These are the gravitational force (mg) downward and the upward spring force, which is equal to the spring constant (k) multiplied by the initial stretching of the spring (y_0). Since the system is in equilibrium at this time, these two forces will be equal and opposite. Thus $mg = ky_0$.

Now consider the moment when the mass is below the base line at distance y but moving upward.

- The gravitational force is downward and equals (–mg).
- The initial (steady state) force exerted by the spring will be upward and equal to $(+ k y_0)$.
- The spring force due to the additional stretch y will be upward and equal to (–ky). At first, it would appear that the negative sign is an error. However, at the selected point in the motion of the mass, y has a negative value, so that the product of y and k would indicate a negative (downward) force, which would be incorrect. The negative

sign in front of the product ky is required to compensate for the negative value of y.

- The force applied by the dashpot is always opposite in direction to the motion. If it were not, there would be no braking action. Since the mass is moving upward, the resistance force will be downward, and equal to

$$-r\frac{dy}{dt}.$$

All of these forces in combination produce the acceleration of the mass. Therefore,

$$m\frac{d^2y}{dt^2} = -mg + ky_0 - ky - r\frac{dy}{dt}.$$

Since mg and ky0 are equal, the differential equation becomes

$$m\frac{d^2y}{dt^2} + r\frac{dy}{dt} + ky = 0.$$

This differential equation describes the motion of a mass suspended from a spring, with damping present. In DC electrical circuitry, there is an equivalent differential equation

$$L\frac{d^2q}{dt^2} + R\frac{dq}{dt} + \frac{1}{C}q = 0,$$

where L, R, and C are the inductance, resistance, and capacitance of the circuit, respectively, and q is the electrical charge.

If the differential equation is written

$$\frac{d^2y}{dt^2} + \frac{r}{m}\frac{dy}{dt} + \frac{k}{m}y = 0,$$

then referring back to the general solution already worked out,

$$A = \frac{r}{m}, \text{ and } B = \frac{k}{m}.$$

$$\text{Then, } \alpha = -\frac{A}{2} = -\frac{1}{2}\frac{r}{m}, \text{ and}$$

$$\omega = \frac{1}{2}\sqrt{4B - A^2} = \frac{1}{2}\sqrt{4\frac{k}{m} - \frac{r^2}{m^2}} = \sqrt{\frac{4km - r^2}{4m^2}} = \sqrt{\frac{k}{m} - \left(\frac{r}{2m}\right)^2}.$$

The end result is

$$y = Ye^{-\frac{r}{2m}t}\sin\left[\left(\sqrt{\frac{k}{m} - \left(\frac{r}{2m}\right)^2}\right)t + \phi\right].$$

What the solution reveals is that the initiating disturbance would cause a displacement Y of the mass. The following displacement y is then modified over a period of time by the cyclic sine function and the damping exponential function. The ϕ term identifies where the mass is in its cycle when t = 0. For the oscillations to begin, the mass must be given an initial displacement downward. If y = –Y when t = 0, substituting these values in the expression for y results in

$$-Y = Y \times 1 \times \sin(0 + \phi).$$

Thus, $\sin\phi = -1$ and $\phi = \frac{3\pi}{2}$ (the low point in the oscillation).

Figure 6 – 2 on the following page is a plot of the mass position y with time for

$$\frac{r}{m} = 0.4, \frac{k}{m} = 12.6, Y = 1.5, \text{ and } \phi = \frac{3\pi}{2}.$$

Units

It is advisable to verify that the units of the factors in the expression have turned out to be correct, bearing in mind that the exponent and argument of the exponential and the sine function are required to be dimensionless. In both cases, the exponent and argument are factors multiplied by t, which has units of seconds. Therefore, these factors should have units of frequency, or the inverse of seconds (per second).

The units of mass are kg. (Note: Within the units the character m stands for metres).

The units of k are force per unit displacement or $kg\frac{m}{s^2}\frac{1}{m} = \frac{kg}{s^2}$.

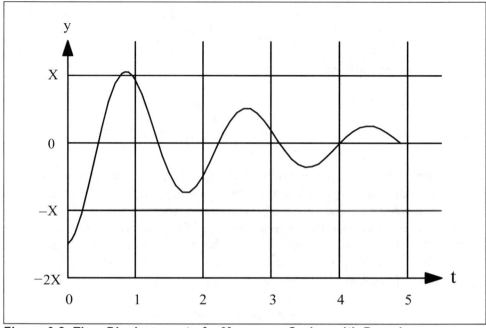

Figure 6-2. Time Displacement of a Mass on a Spring with Damping.

The units of r are force per unit velocity or $kg\dfrac{m}{s^2}\dfrac{1}{\frac{m}{s}} = \dfrac{kg}{s}$.

The exponent of e is $-\dfrac{r}{2m}$. The units are $\dfrac{kg}{s}\dfrac{1}{kg} = \dfrac{1}{s}$, which is correct.

The units of $\dfrac{r}{m}$ are also correct for the argument of the sine function.

While $\left(\dfrac{r}{m}\right)^2$ is involved, it is under the square root sign.

The other expression under the square root sign is $\dfrac{k}{m}$, which will have

units of $\dfrac{kg}{s^2}\dfrac{1}{kg}$ or $\dfrac{1}{s^2}$, which is correct considering the square root function.

A final observation is that if the oscillations of the spring were not damped, then r would be zero. The differential equation would be

$$\frac{d^2y}{dt^2} + \frac{k}{m}y = 0$$

and the solution would be

$$y = Y\sin\left(\sqrt{\frac{k}{m}} + \phi\right)t.$$

Partial Differential Equations

The differential equations analyzed so far have been of the type that have only one dependent variable and one independent variable (usually time). Equations of this degree of complexity are generally adequate for describing the oscillatory behavior which occurs in control systems.

In the real world, however, there are systems in which a single dependent variable may be influenced by more than one independent variable. These systems have to be described by partial differential equations. They may very well apply in operating plants since, for example, analyzing how heat is transferred often requires the use of partial differential equations.

A prominent phenomenon in physics is wave motion. Wave motion in an outward direction occurs when a stone is dropped into a pond. Sound waves travel outward when a bell is truck. Electromagnetic waves travel outward from a radio antenna.

The magnitude of the wave, which is the dependent variable, depends on the point of observation relative to the source of the wave, the direction having three components, and time. The wave equation, familiar to physicists, has the form

$$\frac{\partial^2 u}{\partial x^2} + \frac{\partial^2 u}{\partial y^2} + \frac{\partial^2 u}{\partial z^2} = \frac{1}{c^2}\frac{\partial^2 u}{\partial t^2}.$$

In this equation, u is the dependent variable, that is, the local magnitude of the wave in whatever form it exists; x, y, and z are the positions outward along each of the axes of a 3 dimensional system, c is a constant with units of velocity; and t is time.

A partial derivative, $\frac{\partial u}{\partial t}$ for example, means that when the derivative of u with respect to t is determined, all of the other independent variables involved (in this case x, y, and z) are considered to be constants.

We will likely agree that the solution of the ordinary differential equation, which approximates the behavior of a control system reasonably well, can be sufficiently tedious mathematically without having to resort to the resolution of partial differential equations.

7

Laplace Transforms

History

The origin of Laplace transforms dates back to the era of a British civil engineer named Oliver Heaviside, who lived from 1850 to 1925. An accomplished mathematician, Heaviside was experimenting with what came to be called mathematical operators. In operational calculus, for example, the letter p might replace the operation of taking a derivative, so that if x = f(t), then px was equivalent to

$$\frac{dx}{dt}.$$

Oliver Heaviside's contemporaries ridiculed his work, not so much because of the operators themselves, but because he actually moved operators around in algebraic expressions as if they were ordinary terms. This did not deter him, however, because as far as he was concerned, the method worked.

Eventually, results prevailed, and from Oliver Heaviside's beginnings, mathematicians developed a set of operational transforms that came to be known as the *Laplace transforms*. Just as the use of logarithms can reduce multiplication and division to addition and subtraction, Laplace transforms, where they can be applied, can reduce the problem of solving a differential equation to one of solving an algebraic equation.

In the analysis of control systems, process variables vary with time. The observed behavior may be described by a differential equation, which has time as the independent variable. In such cases, what is required is an expression x = f(t), which describes the behavior of the dependent variable on a time basis, and which is clear of any derivatives. Laplace transforms is a mathematical technique through which this may be achieved.

In Laplace transforms, time is replaced as the independent variable by a new variable, which has been designated s. This has been done by multiplying the function f(t) by e^{-st}, and then integrating the product between the lower and upper limits of zero and infinity. The resulting integral appears as

$$\int_0^\infty e^{-st} f(t)\,dt.$$

The values that s can acquire must necessarily be limited to those that will cause the integral to assume some finite value. The types of differential equations with which this text deals are essentially linear, which means that s will have a real value and will be greater than zero. Most functions f(t) that occur in control systems engineering are Laplace transformable, since with s real and positive, e^{-st} decreases rapidly as t approaches infinity, which more than compensates for an increasing value of f(t).

Inasmuch as the domain of the integral is from zero to infinity, all activity starts at t = 0 and proceeds in the positive direction. Negative values of t have no meaning.

There are two further restrictions on the application of Laplace transforms.

1.   The function f(t) must be single valued, that is, any value of t greater than zero produces only one value of the dependent variable.

2.   More must be known about the relation x = f(t) than just its differential equation. Specifically, the value of f(t) at t = 0 must be known if the differential equation is of the first order. If the differential equation is of the second order, then the value of the first derivative of f(t), that is, the rate of change of the dependent variable must be known at t = 0 as well.

The Laplace transform of a function f(t), denoted L {f(t)} in mathematical shorthand, is defined as

$$L\{f(t)\} = \int_0^\infty f(t)e^{-st}\,dt \text{ and is generally denoted F(s).}$$

If these conditions are satisfied, then the function f(t) is *Laplace transformable.*

Example 1: Step Change

It is worthwhile to work out the Laplace transform for one particular function f(t), since this function becomes important in the study of transfer functions. Fortunately, it is the easiest one to evaluate.

In the process of testing control system components to determine their static and dynamic properties, the input that is used most often is a step input, that is, a sudden jump at t equals zero from a zero signal to a constant signal of finite value. Note that this conforms to the requirement that the initial value of the function be zero. If the input signal suddenly assumes a finite value C at t = 0, then f(t) = C is the function to be transformed. The Laplace transform for the step change will then be,

$$F(s) = \int_0^\infty C e^{-st}\, dt = C \int_0^\infty e^{-st}\, dt = C\left[\frac{1}{-s}e^{-st}\right]_0^\infty$$

$$= C\left[0 - \left(\frac{1}{-s}\right)\right] = \frac{C}{s}.$$

Transforms of Derivatives

The following shorthand symbols are generally used when working with Laplace transforms.

The finite value that x = f(t) assumes at t = 0 is designated x_0. In other words, $x_0 = f(0)$.

The value of $\frac{dx}{dt}$ at t = 0 is designated $\left(\frac{dx}{dt}\right)_0$.

Without going through the mathematical chores involved, we have the following relationships.

1.   The Laplace transform of the function f(t) is designated F(s).

2.   The Laplace transform of the first derivative $\frac{dx}{dt}$ will be $sF(s) - x_0$.

3.   The Laplace transform of the second derivative $\frac{d^2x}{dt^2}$ will be

$$s^2F(s) - sx_0 - \left(\frac{dx}{dt}\right)_0.$$

These facts will be required for the solution of differential equations using Laplace transforms.

Example 2: Time Constant

Figure 7–1 describes the type of dynamic response that is typical of many elements that appear in control systems. The primary characteristic of this component is that at any point on its response curve, the rate of change of the dependent variable with time is proportional to the distance remaining for it to attain its ultimate value. In this case, the dependent variable is y, and in response to a step change input, y begins to change from zero to eventually reach a new value Y.

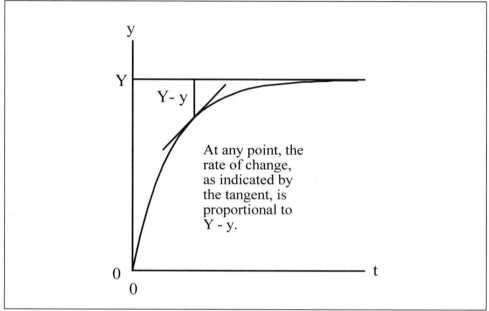

Figure 7-1. Behavior of a Time Constant Element.

At any point in the response of the variable y, the distance remaining is $(Y - y)$. Describing the behavior as a differential equation,

$$\frac{dy}{dt} \alpha (Y - y), \text{ or } \frac{dy}{dt} = \frac{1}{T}(Y - y)$$

where T is a constant. In this expression, the units turn out to be more realistic if $1/T$ is chosen as the constant rather than T.

Rearranging, $T\dfrac{dy}{dt} + y = Y$.

Then, applying the Laplace transformation term by term,

$$T\{s F(s) - y_0\} + F(s) = \frac{Y}{s}.$$

Since $y_0 = 0$, the expression becomes

$$F(s)(Ts + 1) = \frac{Y}{s}, \text{ and } F(s) = Y\frac{1}{s(Ts + 1)}.$$

A table of Laplace transforms reveals that the transform

$$F(s) = \frac{1}{s(Ts + 1)}$$

originates from

$$f(t) = 1 - e^{-\frac{1}{T}t}.$$

Accordingly, the solution to the differential equation,

$$\frac{dy}{dt} = \frac{1}{T}(Y - y) \text{ is } y = Y\left(1 - e^{-\frac{1}{T}t}\right).$$

Since the exponent of e must be dimensionless, and t is in time units, T must also be in units of time. T is, in fact, the time constant of the component.

Example 3: Pendulum

Figure 7–2 shows a pendulum of length L, which oscillates about its fixed center O. At any time t, the angle that the shaft of the pendulum makes with the vertical is θ. Once the free end of the pendulum has been raised so that the pendulum starts from a position θ_0, the pendulum will begin to swing under the influence of gravity.

If we can assume that the mass of the pendulum is m, and that the mass is mainly concentrated at the center of gravity of the bob at the free end of the pendulum, then the force of gravity will be mg downward from its center of gravity.

The path BP is perpendicular to the shaft of the pendulum and is the instantaneous path along which the bob is moving. The angle APB is actually equal to the angle θ. Therefore the component of mg in the direction BP is equal to

mg × the cosine of the angle that BP makes with the vertical

$$= mg \times \cos(90 - \theta) = mg(\cos 90 \cos\theta + \sin 90 \sin\theta)$$

$$= mg(0 + \sin\theta) = mg\sin\theta.$$

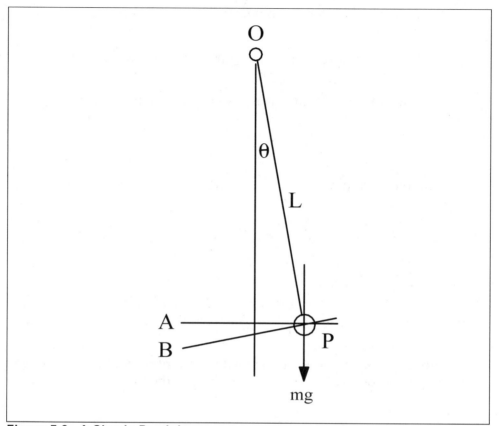

Figure 7-2. A Simple Pendulum.

If the pendulum is made to swing so that the angle θ is kept small, then the value of sin θ becomes virtually the same as the value of θ. Consequently, the component of the gravitational force mg in the direction of motion of the bob will have a magnitude mgθ. However, if we agree that values of the angle θ, and of forces, are positive to the right side of the vertical and negative to the left, then the tangential force exerted on the bob will be $(-mg\theta)$.

The force $-mg\theta$ will be equal to the mass m of the bob multiplied by its acceleration along the path BP. If distances along the arc are denoted by the variable z, then

$$-mg\theta = m\frac{d^2z}{dt^2}.$$

However, we are not so much interested in the position of the weight in its track, with time, as we are in the angle that the shaft makes with the vertical.

The basic relation (arc) z = (radius) L × θ can be differentiated twice, which gives

$$\frac{d^2z}{dt^2} = L\frac{d^2\theta}{dt^2}.$$

$$\therefore -mg\theta = mL\frac{d^2\theta}{dt^2}, \text{ or } \frac{d^2\theta}{dt^2} = -\frac{g}{L}\theta, \text{ and } \frac{d^2\theta}{dt^2} + \frac{g}{L}\theta = 0.$$

What remains now is the solution of this differential equation using Laplace transforms.

Inasmuch as there is a second order derivative involved, it is necessary to know the starting (t = 0) values of θ and of $\frac{d\theta}{dt}$.

Since the motion of the pendulum is cyclical, the starting point can be chosen at any point in the cycle. From the viewpoint of knowing the required starting point values, the best point from which to start is shown in Figure 7-2. With the pendulum in this position, the initial value of θ will be whatever angle is given to the pendulum to start it off. This will be the eventual amplitude of the pendulum and can be designated θ_0.

The angular velocity of the pendulum, $\frac{d\theta}{dt}$, will be zero at this point.

Applying the Laplace transformations term by term,

$$\left\{s^2 F(s) - s\theta_0 - 0\right\} + \frac{g}{L}F(s) = 0,$$

and

$$F(s) = \frac{s\theta_0}{s^2 + \frac{g}{L}} = \theta_0\left(\frac{s}{s^2 + \frac{g}{L}}\right).$$

The table of Laplace transforms shows that the transform $\frac{s}{s^2 + \omega^2}$

originates from the function f(t) − cos ωt. In this case, $\omega^2 - \frac{g}{L}$

and the function sought is consequently

$$\theta = \theta_0 \cos\sqrt{\frac{g}{L}}t.$$

This result is not really as relevant as the period of oscillation of the pendulum. The ω term in the cosine expression has to have units of seconds^{-1} so that the argument ωt is dimensionless. The ω term is actually the angular velocity of the shaft, with units in radians per second.

Since $\omega = 2\pi f$, the frequency f will be equal to $\frac{1}{2\pi}\omega = \frac{1}{2\pi}\sqrt{\frac{g}{L}}$.

The period of oscillation $P = \frac{1}{f} = 2\pi\sqrt{\frac{L}{g}}$.

Example 4: Second Order Linear Differential Equations with Constant Coefficients

Example 2 in Chapter 6 (Differential Equations) dealt with the oscillations of a weight suspended on a spring. The resulting differential equation was

$$\frac{d^2 y}{dt^2} + A\frac{dy}{dt} + By = 0$$

where A and B are constants.

In the spring/mass problem worked as Example 2 in Chapter 6, A stood for r/m, while B stood for k/m. The k term was the constant of the spring, which induces the oscillations. The r term was the damping effect, which tends to reduce or eliminate the oscillations, while the m term was the mass of the weight.

However, it is desirable that the differential equation represents more real world systems than simply that of a weight dangling on a spring. Consequently, the B term in the equation will now be regarded as the "driving force" behind the oscillations, and the A term as the "damping effect." What these two parameters stand for should be kept in mind. The relative magnitudes of the parameters A and B will have a significant impact on the behavior of the system.

The differential equation is linear with constant coefficients, and furthermore, the right side is zero. Accordingly, the resolution of this equation by standard techniques is relatively straightforward, as was shown in Chapter 6. What should now be attempted is the resolution of this differential equation by using Laplace transforms.

Applying the Laplace transforms term by term,

$$\left\{ s^2 F(s) - sf(0) - \frac{df}{dt}(0) \right\} + \left\{ A\left(sF(s) - f(0) \right) \right\} + BF(s) = 0.$$

To continue, it is necessary to declare the initial conditions. Since there is no way of knowing what the velocity of the weight will be when the weight is moving, we are obliged to designate t = 0 at some point where the velocity of the weight is zero. This means at one end or the other of its travel. If the weight is oscillating up and down across the y = 0 base line, then we can specify that it starts out at t = 0 with zero velocity from a position Y below the base line.

With these initial conditions, the transform equation becomes

$$\{s^2F(s) - sY - 0\} + \{AsF(s) - AY\} + BF(s) = 0.$$

Collecting F(s) and Y terms: $F(s)[s^2 + As + B] - Y[s + A] = 0$

which leads to $F(s) = Y\left(\dfrac{s + A}{s^2 + As + B}\right)$.

The procedure now calls for consulting a table of Laplace transforms to obtain the function f(t), which has the expression F(s) determined above as its transform. Unfortunately, the transform

$$\frac{s + A}{s^2 + As + B}$$

does not appear anywhere in the table. Is the conclusion, therefore, that this problem cannot be resolved using the Laplace transform technique?

In fact, Laplace transforms can produce the answer, but here again, the difference between success and failure is the required mathematical experience. The denominator of the transform can be rearranged to complete the square of the first two terms.

$$s^2 + As + B = \left(s^2 + As + \frac{A^2}{4}\right) - \frac{A^2}{4} + B = \left(s + \frac{A}{2}\right)^2 + \left(B - \frac{A^2}{4}\right)$$

There are now three possibilities to be considered. The first is the case in which

$$\frac{A^2}{4} = B.$$

This would imply that the driving force and the damping effect balance each other off. With

$$B - \frac{A^2}{4} = 0$$

the transform becomes

$$\frac{s + A}{\left(s + \dfrac{A}{2}\right)^2}$$

which can be rearranged as

$$\frac{s + \dfrac{A}{2} + \dfrac{A}{2}}{\left(s + \dfrac{A}{2}\right)^2} = \frac{s + \dfrac{A}{2}}{\left(s + \dfrac{A}{2}\right)^2} + \frac{\dfrac{A}{2}}{\left(s + \dfrac{A}{2}\right)^2}$$

$$= \frac{1}{s + \dfrac{A}{2}} + \frac{A}{2} \frac{1}{\left(s + \dfrac{A}{2}\right)^2}.$$

An important fact of mathematics, one which is not disclosed in many texts on Laplace transforms, now emerges: *If a Laplace transform, which cannot be inverted as it stands, can be expressed as the sum of two parts, each of which is capable of being inverted, then the solution will be the sum of the inverted parts.*

From the table of Laplace transforms, the inverse transforms (in mathematics shorthand, L^{-1}) for this problem are

$$L^{-1}\left(\frac{1}{s + \dfrac{A}{2}}\right) = e^{-\frac{A}{2}t}, \text{ and } L^{-1}\frac{1}{\left(s + \dfrac{A}{2}\right)^2} = t\, e^{-\frac{A}{2}t}$$

$$\therefore \text{ the solution } f(t) = Y\left(e^{-\frac{A}{2}t} + t\, e^{-\frac{A}{2}t}\right) = Y\, e^{-\frac{A}{2}t}\left(1 + t\right).$$

Since the solution does not contain either a sine or cosine function, the system is not going to oscillate. Furthermore, this system had as its basis that

$$\frac{A^2}{4} = B.$$

Note that if $\dfrac{A^2}{4}$ were greater than B, then the damping effect that A represents would be even more dominant.

Consequently, for any value of $\dfrac{A^2}{4}$ that is equal to or greater than B, the system will not oscillate.

The Oscillatory Case

The remaining possibility, therefore is the one in which B is greater than $\dfrac{A^2}{4}$, and the driving force is dominant over the damping effect.

The Laplace transform $F(s) = Y\left(\dfrac{s + A}{s^2 + As + B}\right)$

$$= Y\left[\dfrac{s + A}{\left(s + \dfrac{A}{2}\right)^2 + \left(B - \dfrac{A^2}{4}\right)}\right].$$

To simplify the various expressions, let $A/2 = a$, and let

$$\omega^2 = \left(B - \dfrac{A^2}{4}\right).$$

Since B is greater than $A^2/4$, ω^2 will be a positive quantity. Later it will be seen that using ω^2 instead of ω will make it easier to invert the transform.

Accordingly, $F(s) = Y\left(\dfrac{s + \dfrac{A}{2} + \dfrac{A}{2}}{\left(s + \dfrac{A}{2}\right)^2 + \left(B - \dfrac{A^2}{4}\right)}\right)$

$$= Y\left(\dfrac{s + a}{(s + a)^2 + \omega^2} + \dfrac{a}{(s + a)^2 + \omega^2}\right)$$

$$= Y\left(\dfrac{s + a}{(s + a)^2 + \omega^2} + \dfrac{a}{\omega}\dfrac{\omega}{(s + a)^2 + \omega^2}\right).$$

From the table of Laplace transforms:

$$L^{-1}\left(\frac{s+a}{(s+a)^2+\omega^2}\right) = e^{-at}\cos\omega t, \text{ and } L^{-1}\left(\frac{\omega}{(s+a)^2+\omega^2}\right) = e^{-at}\sin\omega t$$

$$\therefore \text{the solution is } y = f(t) = Y\left[e^{-at}\left(\frac{a}{\omega}\sin\omega t + \cos\omega t\right)\right]$$

$$= Ye^{-\frac{A}{2}t}\left(\frac{A}{2\omega}\sin\omega t + \cos\omega t\right),$$

where $\omega = \sqrt{B - \frac{A^2}{4}}$, or $\frac{1}{2}\sqrt{4B - A^2}$ as in Chapter 6.

Consistency of Results

An astute observer would notice that the solution obtained through the use of Laplace transforms is not the same as that obtained in Chapter 6 for the second order linear differential equation with constant coefficients, which was

$$x = Xe^{\alpha t}\sin(\omega t + \phi) \quad (\text{ where } \alpha \text{ was substituted for } -A/2).$$

In this expression, x is a general variable, not necessarily the displacement of a mass on a spring. X will be the original amplitude of the oscillations, whereas M, where it appears, is the value of x at t = 0.

When the differential equation was solved using the Laplace transforms, it was necessary to specify two initial conditions, and this fact must be recognized when comparing the two apparently different results. Specifically, this means that the initial conditions, which were specified when using the Laplace transforms, should be applied to the solution obtained in Chapter 6.

The initial conditions were:

1.   When t = 0, x = M.

2.   When t = 0, $\frac{dx}{dt} = 0$.

Applying the initial condition (1) to the solution $x = Xe^{\alpha t}\sin(\omega t + \phi)$:

$$M = X \times 1 \times \sin(0 + \phi), \text{ from which } X = \frac{M}{\sin\phi}.$$

To apply initial conditions (2), it will be necessary to take the derivative of x with respect to t for $f(t) = X e^{\alpha t} \sin(\omega t + \phi)$. The rules for taking the derivative of a product, and of a function within a function, are needed (refer to Chapter 2).

$$\frac{d}{dt}\left[X e^{\alpha t} \sin(\omega t + \phi)\right] = X\left[\frac{d}{dt} e^{\alpha t} \times \sin(\omega t + \phi) + e^{\alpha t} \times \frac{d}{dt} \sin(\omega t + \phi)\right]$$

$$= X\left[\alpha e^{\alpha t} \sin(\omega t + \phi) + e^{\alpha t}\cos(\omega t + \phi) \times \omega\right]$$

$$= X e^{\alpha t}\left[\alpha \sin(\omega t + \phi) + \omega\cos(\omega t + \phi)\right] = \frac{dx}{dt}.$$

Since $\frac{dx}{dt} = 0$ when $t = 0$, $X \times 1 \times \left[\alpha \sin(0 + \phi) + \omega\cos(0 + \phi)\right] = 0$ and

$\alpha \sin \phi + \omega\cos \phi = 0$. Rearranging this:

$$\frac{\sin \phi}{\cos \phi} = -\frac{\omega}{\alpha} = \tan \phi.$$

Now, $X e^{\alpha t}\sin(\omega t + \phi) = X e^{\alpha t}\left(\sin \omega t \cos \phi + \cos \omega t \sin \phi\right)$

$$= \frac{M}{\sin \phi} e^{\alpha t}\left(\sin \omega t \cos \phi + \cos \omega t \sin \phi\right)$$

$$= M e^{\alpha t}\left(\sin \omega t \frac{\cos \phi}{\sin \phi} + \cos \omega t \frac{\sin \phi}{\sin \phi}\right)$$

$$= M e^{\alpha t}\left(\sin \omega t \frac{1}{\tan \phi} + \cos \omega t\right) = M e^{\alpha t}\left(\sin \omega t \times \left(-\frac{\alpha}{\omega}\right) + \cos \omega t\right).$$

$$= M e^{\alpha t}\left(\cos \omega t - \frac{\alpha}{\omega}\sin \omega t\right).$$

Finally, since

$$\alpha = -\frac{A}{2}, \quad x = M e^{-\frac{A}{2}t}\left(\frac{A}{2\omega}\sin \omega t + \cos \omega t\right),$$

which is the solution obtained through the use of Laplace transforms.

Table of Laplace Transforms

A table containing some of the more common Laplace transforms is contained in Table 7-1.

Table 7-1. Short Table of Laplace Transforms

Function $x = f(t)$ ($t > 0$)	Laplace Transform $F(s) = L\{f(t)\}$
C (constant)	$\dfrac{C}{s}$
t	$\dfrac{1}{t^2}$
t^2	$\dfrac{2}{s^3}$
t^n	$\dfrac{n!}{s^{n+1}}$
$e^{-\omega t}$	$\dfrac{1}{s + \omega}$
$\dfrac{1}{T} e^{-\frac{t}{T}}$	$\dfrac{1}{Ts + 1}$
$1 - e^{-\frac{t}{T}}$	$\dfrac{1}{s(Ts + 1)}$
$f(t - L)$	$e^{-Ls} F(s)$
$\sin \omega t$	$\dfrac{\omega}{s^2 + \omega^2}$
$\cos \omega t$	$\dfrac{s}{s^2 + \omega^2}$
$e^{-\alpha t} \sin \omega t$	$\dfrac{\omega}{(s + \alpha)^2 + \omega^2}$
$e^{-\alpha t} \cos \omega t$	$\dfrac{s + \alpha}{(s + \alpha)^2 + \omega^2}$
dx/dt	$s F(s) - x_0$
d^2x/dt^2	$s^2 F(s) - s x_0 - (dx/dt)_0$

x_0 and $(dx/dt)_0$ are the values of x and dx/dt when $t = 0$.

8

Frequency Response Analysis

Background

Our primary objective in studying control systems is to understand how they behave. Then through understanding their behavior, hopefully we can exert some influence to cause them to perform in a manner that will be beneficial.

Just observing behavior is not quite enough, however. It is necessary to have definite criteria of behavior, and the criteria must be measurable so that comparisons can be made. Frequency response analysis is one method of meeting these needs.

A frequency response test of a control system component (or even of a whole control system) is conducted by forcing a test signal which varies in sine wave fashion into the input of the component. At the same time, the output of the component is tracked so that the input and the output can be compared. The unique property of the sine wave input is that it is the only type of input that produces an output of the identical form. The output will also have the sine wave shape, and its frequency of oscillation will be the same as that of the input. Hence the name, *frequency response.* This is where the similarity ends, however.

Figure 8-1 is a graph of the frequency response input and output of a component under test. Comparing the output wave with the input, two factors are significant. First, the inherent gain of the component has modified the magnitude of the output wave. In this case it emerges smaller in magnitude than that of the input wave. Second, on the time scale, the output wave is out of phase with the input wave. It actually lags behind the input

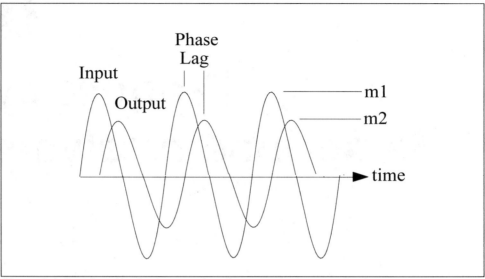

Figure 8-1. Input and output waves in a frequency response test.

wave. This is caused by the component's inherent reaction time. The slower the reaction of the component, the greater will be the time lag.

The ratio of the magnitude of the output wave to that of the input wave is called the *magnitude ratio*. It is measured by the ratio of m_2/m_1 in Figure 8-1. The phase (time) lag of the output wave is measured in degrees. If the output wave trailed the input by one whole cycle, then the phase lag would be 360°. The phase lag shown would be about 1/4 cycle or 90°.

The graph shows the magnitude and the phase lag at only one frequency. In a complete analysis, the component would be subjected to a sine wave input over a whole range of frequencies, with the magnitude ratio and phase lag being measured at each one. The variation of the magnitude ratio and the phase lag over the relevant range of frequencies are the two performance criteria that frequency response analysis yields.

The Bode Diagram

Once the magnitude ratio and phase lag data have been accumulated for the range of frequencies of interest, it is customary to plot both of these data on a base of frequency. The graph in Figure 8-2 is an example. Graphs with the frequency response data made visible in this form are often called Bode diagramsgain, after H.W. Bode, who was a noted pioneer in the development of the theory of feedback amplifiers. While process control systems function at considerably lower frequencies than those with which Bode would have been dealing, his manner of presenting the data is nevertheless applicable.

In the Bode diagram, the magnitude ratio of the output to the input is generally abbreviated as *gain*. The frequency and gain scales in the diagram are logarithmic, so they can cover a range of two or more decades in a reasonable space. Setting out the scales linearly would spread the diagram out to the point where it would be unwieldy.

Frequencies are measured in cycles per minute (cpm); the practical unit considering the slow rate of the oscillations. A commonly used scale is from 0.01 to 1 cpm, as shown in Figure 8-2. In process control, frequency rates are so low that control theorists usually talk in terms of the period of oscillation, which is the time required to make one cycle, rather than frequency. The period of oscillation will be 1/frequency.

Frequency Response of a Time Constant Element

The frequency response data in Figure 8-2 is for an element that occurs in all process control systems, namely, a time constant. The value of the time constant in this case is 2 minutes. This element also has a steady state gain of 5.0. In frequency response analysis, steady state is equivalent to zero frequency. Logarithmic scales, by their nature, cannot go down to zero, but a value of 0.01 cpm is usually low enough to reveal the steady state gain of the component under test.

The steady state gain of 5.0 means that at very low frequencies the amplitude of the output wave will be 5 times that of the amplitude of the input wave. The gain graph shows that even with a time constant of 2 minutes, which implies a relatively slow reaction to any input, at the very low frequencies the output of the element still tracks the input, and the gain of 5 is maintained. As the frequency increases, however, a point is reached where the output wave is not completed before a new wave arrives at the input of the element. From then on, the amplitude of the output wave decreases until at a frequency of 1 cpm, the gain has dropped to 0.4.

The phase lag graph shows that even at the minimum frequency of 0.01 cpm, the output wave lags slightly behind the input wave. From then on, the phase lag increases with the frequency. Interestingly, even at a theoretical infinite frequency, the phase lag never exceeds –90°. This is a unique property of a time constant element.

Frequency Response of a Dead Time Element

A frequency response plot for a dead time element is shown in Figure 8-3. Graphs are shown for three different levels of dead time: 0.2, 0.5, and 1 minute. In process control studies, dead time elements are considered to contribute phase lag to the system but no gain. Thus the gain graph for

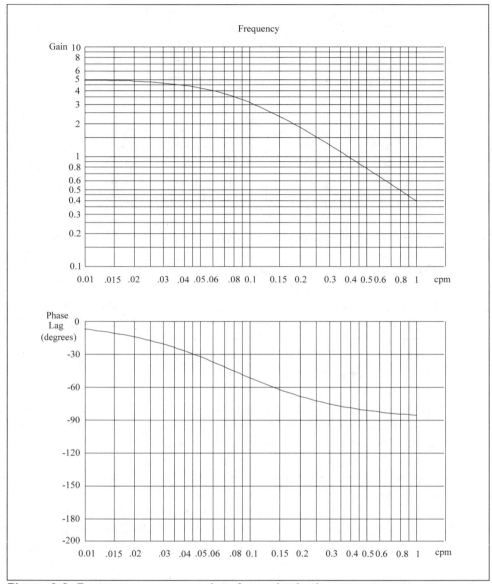

Figure 8-2. Frequency response data for a single time constant.

each of the three dead time elements is 1.0 across the whole range of frequencies.

The damage that dead time creates in feedback control systems is shown in the rapid way in which dead time increases phase lag as the frequency increases. This can have a harmful effect on the time required for the control system to recover after it is disturbed. Further discussion of this important point will follow later.

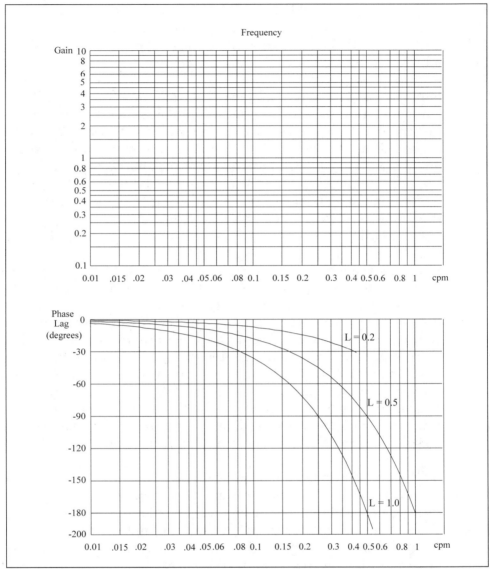

Figure 8-3. Frequency response data for assorted dead time elements.

Notice that at a frequency of 0.25 cpm, the phase lag contributed by a 1 minute dead time element is –90°. For the 0.2 and 0.5 minute dead time elements it is –18° and –45°, respectively, at the same frequency. This shows clearly that the phase lag created is in direct proportion to the dead time that is present.

Combinations of Components

If the frequency response data for the individual components are available, then the frequency response characteristics that two or more components in tandem would have can be easily computed. It is only necessary to *multiply* the individual *gain* values, and *add* the individual *phase lag* values, taken at the selected frequency. The rule: gains multiply, phase lags add.

Suppose that a system consisted of a 0.5 minute dead time, followed by a 2 minute time constant. The gain and phase lag of the combination can be determined at any frequency from the graphs in Figure 8-2 and 8-3. From Figure 8-2, at a frequency of 0.5 cpm, the gain for the time constant is 0.78, and the phase lag is –81°. From Figure 8-3, at the same frequency, the gain for the 0.5 minute dead time element is 1.0, and the phase lag is –90°. Accordingly, at a frequency of 0.5 cpm, the gain of the time constant and dead time together will be 0.78 x 1.0 = 0.78, and the phase lag will be –81° + (–90°) = –171°.

It follows that if the frequency response characteristics of the automatic controller, the control valve, the process, and the measurement sensor, which are the essential components of the control system, were available, then the frequency response characteristics of the whole system could be computed, and from that, the ultimate performance of the control system on control predicted. During the 1960s, there was a definite impetus to predict control system performance in this way. Unfortunately, the procedure requires that the characteristics of all of the components be known, and while it was not difficult to obtain this data for the controller, the control valve, and the sensor, the data for the dominant component—the process that was to be controlled—was always lacking.

Since there was no telling theoretically what mixture of dynamic elements the process might have in it, the only alternative was to obtain the data by making an actual field test. This would involve using a special signal generator to disturb the process in sinusoidal fashion, over a whole range of frequencies, and recording what resulted at the process's output. No process operator in his right mind who was in charge of a boiler, a fractionation column, or a reactor, would allow such a test to be made.

This obstacle would have brought about the demise of frequency response analysis as it applies to process control systems, had it not been for the contribution of J.G. Ziegler, who developed another more realistic method of determining the characteristics of a process. This is discussed in Chapter 10.

Period of Oscillation

When part of the output of any system is fed back into its input, then a closed loop is created, which effectively sets the stage for oscillations to occur. Anyone who has pointed a microphone at a loudspeaker knows this. Of the various performance criteria that apply to control systems, the time required to make one oscillation has the greatest impact. This time value is called the *period of oscillation*. It is actually the inverse of the frequency of oscillation that is the basis for plotting gain and phase lag in frequency response diagrams.

The period of oscillation of a control system is an inherent property of the system. It is created by the dynamic characteristics of all of the components in the system in combination. As such, in real life process control systems the period of oscillation cannot be appreciably altered. We have to live with it.

If a feedback control system is disturbed, the automatic controller usually does not recognize the upset in the process and its effect on the controlled variable until the measurement sensor has actually measured a change in the controlled variable and has fed this information back to the controller. Owing to the response times of the process and the sensor, corrective action by the controller does not take place until some time after the disturbance has occurred; in other words, it happens too late. This means that after each disturbance, the control system has to go through an interval of upset and recovery before it can get back on control.

When the controller is correctly tuned to provide the right amount of corrective action and to apply it no faster than the process can absorb it, the pattern of the recovery will be a sine wave with each peak smaller than the peak that preceded it, until the oscillations die out altogether. Figure 8-4 illustrates this.

Adjusting the controller for a good recovery in a feedback control system is a compromise between minimizing the height of the first peak, as that is when the controlled variable deviates farthest from the desired value, and minimizing the recovery time, which is the time required to get back on control. Most control experts agree that this compromise is best achieved if the recovery curve exhibits oscillations of decreasing amplitude, with the amplitude of each peak being about 1/4 of the amplitude of the peak that preceded it, as shown in Figure 8-4. The important feature that Figure 8-4 illustrates is that if the control system recovers in the optimum manner, then following a disturbance it usually takes two or three oscillations to get back on control. This being so, then the time required to make one oscillation, that is, the period of oscillation, becomes all important.

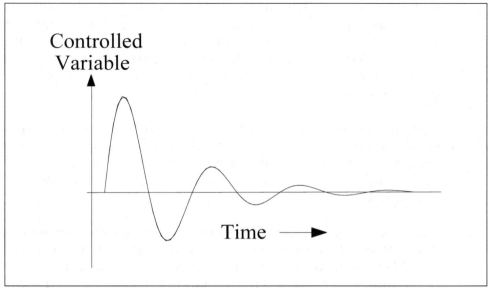

Figure 8-4. Control system recovery at an amplitude ratio of ¼.

Summary

It may well be, therefore, that the most valuable item of information that the frequency response diagram can give us is the value of the period of oscillation created by the composite of components which comprise the control system.

If conditions are favorable, or in another sense, unfavorable, a feedback control system can go into a state of continuous oscillations. Chapter 9 describes the conditions that must prevail to cause this. Chapter 9 also explains the fact that when a control system oscillates, it will do so at the frequency at which the cumulative phase lag of all the components of the system becomes $-180°$.

Thus, if the frequency response gain and phase lag can be plotted for the complete system of components, then the frequency at which the phase lag curve crosses $-180°$ will be the frequency at which the control system is going to oscillate. The period of oscillation will be the inverse of this frequency, and the recovery time of the control system following a disturbance will be two or three times the period of oscillation. It is necessary to settle for a ballpark factor of two or three times, rather than a definite number, since the recovery time will also depend on the size of the disturbance. A more severe disturbance may result in a greater number of oscillations before the control system settles down.

Frequency response analysis proves, among other things, that the bad control system components are those that contribute excessive phase lag and, consequently, cause the phase lag curve to cross the $-180°$ line at lower frequencies. Lower frequencies mean longer periods of oscillation and longer control system recovery times.

9

Transfer Functions and Block Diagrams

Background

Feedback control systems are made up of components that are reactive by nature. This means that each one has an input (sometimes more than one) and the means to generate an output. The inputs and outputs have a variety of forms, but in process control the most common are process variables and instrument signals.

To be usable, the output of a component must exhibit a consistent relationship to its input. Output relationships are not necessarily neat and tidy, but the same input must consistently produce the same output; otherwise the component is unacceptable.

Control systems experts need techniques to determine and describe how the components of a control system will perform. If the behavior of the individual components that make up the system can be identified, then the behavior of the overall system can be evaluated.

This leads to the question: What kind of behavior are we interested in? There are two factors:

1. The gain of the component. If the input to the component is changed a known amount, how much does its output change? The gain factor will be the ratio of the change in output to the change in the input that created it. The output change is measured as it goes from the initial steady state value to the final steady state value. Time is not a factor. The output is permitted all the time that is necessary to assume its new value.

A further consideration is whether or not the gain of the component remains the same over the whole range of operation of the component. If it doesn't, then this factor has to be allowed for in the control system design.

2.   The dynamic response of the component. This does not mean how much the output of the component responds, but how quickly and on what pattern it assumes its new steady state output after its input is changed. Components that react quickly make for better overall control system performance.

Transfer Functions

While frequency response analysis uses a graphical method to describe the gain and dynamic response of components, transfer functions do the same thing using mathematical expressions. By using transfer functions, it is often possible to describe both the gain and the dynamic response of a component in a single mathematical function.

In general, the transfer function of a component is the ratio of the change in its output to the change in its input, but herein lies a problem. While the gain of the component can be identified by a simple number, the dynamic character (how it varies with time) of both the input and output can be described only by differential equations in which time is the independent variable, and then only if the input or output function is continuous, which it may not be. Obviously, a transfer function that consisted of the ratio of two differential equations would be of little practical use. The situation can be made workable, however, not by using the differential equations of the input and output, but by using their Laplace transforms.

The bottom line is that in feedback control systems, the transfer function of a component is defined as the ratio of the Laplace transform of its output to the Laplace transform of its input. What is now required are some examples of transfer functions, and then a study of just what performance information can be gained from them.

The Step Input Function

Frequency response analysis of a component or system is based on a sine wave input signal. A sine wave input results in a sine wave output. It is obvious, however, that if the input were anything other than a sine wave, then the output would be different, even though it emerged from the same component. For transfer functions to have practical value, therefore, it was necessary to standardize the form of the input signal.

In field test work, the most practical test signal is a step change from one input level to another, for at least two reasons. First, it is an easy test signal to devise, and second, the output that results from this relatively simple input change will yield all of the dynamic information that is of any real value. It is not surprising, therefore, that the step change was selected as the standard input for transfer functions. A further refinement was that the step input should have a unit value.

In the chapter on Laplace transforms, it was shown that the transfer function F(s), which results from a step change from zero to a value of C, is F (s) = C/s. If C = 1, then F(s) becomes 1/s. The simplicity of this function is another plus for an input consisting of a unit step change.

In real life process control systems, two particular components, time constants and dead time, predominate over all others. The transfer functions for these two components should now be worked out.

Time Constants

In Chapter 6 (Differential Equations), Example 1 described a component whose rate of change, in response to a step change input, is proportional to the distance remaining for the output to attain its ultimate value. The output in this case is the variable x. In the simplest case, the input and the output have the same value at t = 0, and the gain of the component is 1. Then the equation x = f(t) for the output is:

$$x = \left(1 - e^{-kt}\right).$$

The exponent of the exponential e is required to be dimensionless, and the units of t are time units. Accordingly, it is more realistic to set k = 1/T. T will now be in time units. T is, in fact, the time constant of the component. The differential equation thus becomes

$$x = \left(1 - e^{-\frac{1}{T}t}\right).$$

The table of Laplace transforms shows that the function

$$f(t) = \left(1 - e^{-\frac{1}{T}t}\right)$$

has the transform

$$F(s) = \frac{1}{s(Ts + 1)}.$$

Consequently, the transfer function for a time constant element will be

$$\frac{\text{Laplace transform of the output}}{\text{Laplace transform of the input}} = \frac{\dfrac{1}{s(Ts+1)}}{\dfrac{1}{s}}$$

$$= \frac{1}{Ts+1}.$$

Dead Time

If a component has dead time, then the output function will duplicate the input function f(t), but only after a delay of L time units. L is usually referred to as the *dead time*. The output function will consequently be f(t – L). The initial value of t must be zero (a Laplace transforms requirement), after which t increases positively.

The table of Laplace transforms shows that the transform for f(t – L) is $e^{-Ls} F(s)$, where F(s) is the transform for f(t). Thus, for whatever form f(t) may have, the transfer function for a dead time element will be

$$\frac{\text{Transform of the output}}{\text{Transform of the input}} = \frac{e^{-Ls} F(s)}{F(s)} = e^{-Ls}.$$

The Value of the Transfer Function

The goal in developing transfer functions was to devise a mathematical expression that would incorporate both the steady state gain and dynamic characteristics of a control system component. It now remains to be shown that this goal has been achieved.

A useful attribute of the transfer function is that by applying the appropriate procedure, the transfer function will yield the frequency response data of its component. More specifically, from the transfer function, other functions will evolve from which the frequency response magnitude ratio and phase angle can be determined at any desired frequency. The procedure is as follows:

1.   In the transfer function, replace the operator s with $j\omega$, where j is the imaginary quantity $\sqrt{-1}$, and ω is the angular velocity (radians per second).

2.   Using the standard techniques for complex numbers, separate the resulting expression into its real part (RP) and imaginary part (IP). This will lead to the expressions for the frequency response magnitude ratio (MR) and the frequency response phase angle (ϕ). These expressions will be functions of the angular velocity ω, which is directly related to the frequency by the relation $\omega = 2\pi\,f$.

3.   The magnitude ratio at the specified value of ω will then be

$$\text{MR} = \sqrt{(\text{RP})^2 + (\text{IP})^2}\,.$$

4.   The phase angle at the specified value of ω will be

$$\phi = \text{angle whose tangent is } \frac{\text{IP}}{\text{RP}}, \text{ that is, } \phi = \tan^{-1}\!\left(\frac{\text{IP}}{\text{RP}}\right).$$

Example 1: Time Constant

The transfer function for a time constant component is

$$\frac{1}{Ts + 1}.$$

Converting to the frequency response domain, this expression becomes

$$\frac{1}{j\omega T + 1}.$$

To separate the real and imaginary parts:

$$\frac{1}{j\omega T + 1} = \frac{1}{j\omega T + 1} \times \frac{j\omega T - 1}{j\omega T - 1} = \frac{j\omega T - 1}{-\omega^2 T^2 - 1} = \frac{1 - j\omega T}{-\omega^2 T^2 + 1}$$

$$= \frac{1}{\omega^2 T^2 + 1} - j\frac{\omega T}{\omega^2 T^2 + 1}.$$

The real part $\text{RP} = \dfrac{1}{\omega^2 T^2 + 1}.$

The imaginary part $\text{IP} = \dfrac{-\omega T}{\omega^2 T^2 + 1}.$

The magnitude ratio $= \sqrt{(RP)^2 + (IP)^2} = \sqrt{\left(\dfrac{1}{\omega^2 T^2 + 1}\right)^2 + \left(\dfrac{-\omega T}{\omega^2 T^2 + 1}\right)^2}$

$$= \frac{1}{\omega^2 T^2 + 1}\sqrt{1 + \omega^2 T^2} = \frac{1}{\sqrt{\omega^2 T^2 + 1}}.$$

The phase angle $= \tan^{-1}\dfrac{IP}{RP} = \tan^{-1}\dfrac{\left(\dfrac{-\omega T}{\omega^2 T^2 + 1}\right)}{\left(\dfrac{1}{\omega^2 T^2 + 1}\right)} = \tan^{-1}(-\omega T).$

It should be noted that in the development of the magnitude ratio and phase relationships above, the transfer function for the time constant was

$$\frac{1}{Ts + 1}.$$

This would indicate a steady state gain of 1, which is not necessarily the case. If the time constant component contributes a steady state gain (magnitude k), as well as dynamics, then the transfer function would be

$$k\frac{1}{Ts + 1}$$

and the magnitude ratio would be

$$\frac{k}{\sqrt{\omega^2 T^2 + 1}}.$$

Example 2: Dead Time

The transfer function for a dead time element is $F(s) = e^{-Ls}$. In the frequency response domain, this becomes $e^{-j\omega L}$. Applying the work done in Chapter 5 (Complex Quantities), $e^{-j\omega L} = \cos \omega L - j \sin \omega L$.

Therefore, $RP = \cos \omega L$, and $IP = -\sin \omega L$.

The magnitude ratio will be $\sqrt{\cos^2 \omega L + \sin^2 \omega L} = \sqrt{1} = 1$.

In control system analysis, it is considered that there is only one dead time element, if any. If dead time is present in more than one place in a control system, all of the dead times can be summed together to form a single dead time component, without any loss of accuracy in the analysis.

The expression for the magnitude ratio shows that since the magnitude ratio is 1, dead time does not contribute any gain or attenuation to the control system, irrespective of the frequency. Increases or decreases in the overall system gain will be contributed by components of some other type, most likely by time constants.

The phase angle will be $\phi = \tan^{-1}\left(\dfrac{-\sin\omega L}{\cos\omega L}\right) = \tan^{-1}(-\tan\omega L) = -\omega L.$

Block Diagrams

A fact of life that sometimes eludes control system theorists is that process control earns its keep in industry, not in the lab. Furthermore, the brand of process control that is dominant in industry, and which will continue to dominate, is feedback control. There are two reasons for this.

First, feedback control requires the minimum investment. All that is required is a controlling device (analog or digital), a sensor from which the controller can get information about the variable it is supposed to control, and a final device, such as a control valve, which can manipulate some other variable whose value affects the value of the variable under control.

Second, to create a feedback control system, there is no requirement for any extensive engineering study. As far as the process to be controlled is concerned, it is only necessary to know that the same input to the process consistently produces the same reaction from the process.

It helps in the study of what goes on in a feedback control system if the system is diagrammed in block form, with each of the major components in the system represented by one block. The major components in a feedback system are the process, the controller, the sensor from which the controller gets its measurement information about what it is controlling, and the final device, which the controller uses to effect appropriate changes in the controlled variable. In a block diagram, a feedback control system appears in Figure 9-1.

The first (circular) block is the comparison block, in which the measured value of the controlled variable c is compared with the desired value r. The difference, ε, is equal to c − r. If the control system is on control, then $\varepsilon = 0$. If ε is not zero, corrective action by the controller is required. The comparison block is actually built into the controller but is shown separately in the

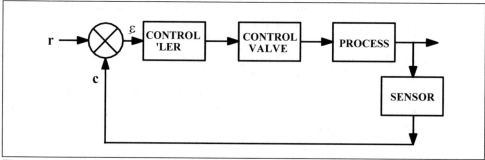

Figure 9-1. The Feedback Control System Block Diagram.

block diagram so that the location of the variables r, c, and ε can be identified.

The block diagram shows how the output of one block becomes the input of the next. Each block may contain one or more time constants, dead time, or other behavioral characteristics. The input to every block will be modified by whatever gain and dynamics the previous component contributes to the system.

The output of the process is the variable that is being controlled. In real life it is more likely to be a process operating condition such as pressure or temperature, rather than an actual product.

The diagram also shows the measured value of the controlled variable being fed back to the input of the controller. Hence the name *feedback control*. In addition, having the feedback path means that the system comprises a closed loop. Accordingly, the term *closed loop control* is also used for systems of this kind.

In any system, when some or all of the output of the system is fed back as an input to the system, this creates conditions under which the system can oscillate. Oscillations do occur in process control systems, and as such, are an important characteristic of the system.

The overall behavior of the control system can be determined if the characteristics of each of the components represented by its block can be established. This may be achieved in either of two ways.

1.   If the output of the component in response to a step change input can be expressed as a differential equation, then a transfer function for the component can be written, and the frequency response magnitude ratio and phase angle data can be worked out.

2.   The steady state and dynamic characteristics of the component can be determined by conducting an appropriate test. This is the method that usually has to be used to determine the characteristic of real processes.

In some texts, the control valve and sensor blocks are lumped into the process block, as in Figure 9-2.

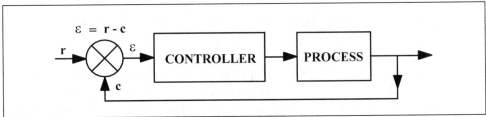

Figure 9-2. The Abbreviated Feedback Control System Block Diagram.

Since the controller does not know how many components are downstream of it, only the results that come back to it, it is considered satisfactory to show all three of these components as a single block.

An apparent anomaly may be present if the expression for the control error ε is written as ($\varepsilon = r - c$). From the logical point of view, the control error should be positive if c is greater than r, but this expression says the opposite. This point requires clarification, which will be forthcoming after one further matter has been considered.

Conditions for Continuous Oscillation

The feedback nature of the control system makes it possible for oscillations to occur. This raises the question: What conditions are necessary if the oscillations in the system are to be self sustaining and therefore continuous? Suppose that by using a signal generator or by some other artificial means, a sine wave input is introduced into the system via the r input. The oscillations would proceed around the loop and back to the comparator block as the c input. If the applied sine wave were then withdrawn, under what conditions would the oscillations continue on their own? Actually, two conditions would have to be met.

1.   The magnitude of the sine wave that returned via the c input would have to be as great as the magnitude of the sine wave that was applied at the r input. If the c input magnitude were less than that of the r input, then the oscillations would die out.

2.   The oscillations returning at the c input must be in phase with the sinusoidal r input. If the magnitude of the c sine wave input were

the same as that of the r input, but the two wave trains were out of phase, then the interference between the two out-of-phase waves would cause the oscillations to die out.

In real process control systems, the output from the process block will inevitably emerge lagging behind the input, owing to the dynamic delays that are inherent in the control valve, the process, and the measurement sensor. Thus, for the returning wave at the c input to be in phase with the incoming wave r, the returning wave must actually be 360° or one complete cycle out of phase with the input r. This effectively puts it back in phase with the r wave.

In Chapter 8 (Frequency Response Analysis), it was shown that the phase lag will increase with increasing frequency of the oscillations. From this it might be deduced that continuous oscillations in the system can occur only at the frequency that causes a phase lag of 360° in the process block.

However, at this point a second anomaly occurs. Control theory texts say that continuous oscillations occur at the frequency at which there will be 180°, not 360°, of phase shift in the process. How can this be reconciled?

The answer lies in the expression $\varepsilon = r - c$. We have already noted that this does not appear to be logical, but the difficulty lies in the fact that the expression has been abbreviated. The actual expression is

$$\varepsilon = (-1) \times (c - r).$$

This clears up the two apparent inconsistencies. First, the difference term is really $(c - r)$, not $(r - c)$, so logic prevails. The second is that multiplying a sine wave by (-1) inverts the wave, which is equivalent to shifting the phase by 180°. This inversion occurs inside the circular block. Consequently, 180° of phase lag in the process, plus the 180° contributed by the inversion factor –1, provides the overall 360° phase lag that is required to make continuous oscillations possible. The key factor to remember is that *continuous oscillation in a feedback control system will occur at the frequency that causes 180° of phase lag in the process.*

It is theoretically possible that all of the dynamic elements in the process block, in combination, will not produce a phase lag of 180°, no matter how great the frequency. In this case, oscillations will not be sustained in the loop, irrespective of the steady state gain in the loop. Such a hypothetical control system would have to have zero dead time and no more than two time constants, even if they are small. Since in actual control systems there will be at least one time constant in each of the control valve, the process, and the sensor, a real process control system that cannot oscillate does not exist.

The Transfer Function of a Closed Loop

The diagram below is the abbreviated closed loop diagram except for the different symbols. The transfer functions are identified by the letter G, so that G_c is the transfer function of the controller, and G_p is the transfer function of the process. The input and output, which will be functions of time, are identified by the letter Z, with Z_i being the input to the system, and Z_o being the output.

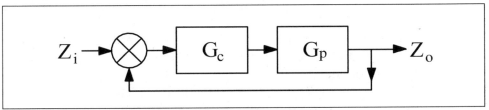

Figure 9-3. Block Diagram with Transfer Functions.

For the two blocks, the input to the second block is the output from the first, so the transfer function for the two blocks in tandem will be the product G_cG_p. At first glance, it seems that the ratio of the system output to the input should be

$$\frac{Z_o}{Z_i} = G_c G_p.$$

This would be the case if there were no feedback. However, the feedback path that directs the output back to the input of the loop, adds a complication. The input to the G_c block will not be Z_i, but $Z_i - Z_o$. Consequently,

$$\frac{Z_o}{Z_i - Z_o} = G_c G_p. \text{ So, } Z_o = G_c G_p Z_i - G_c G_p Z_o$$

$$Z_o \left(1 + G_c G_p\right) = G_c G_p Z_i \text{ and } \frac{Z_o}{Z_i} = \frac{G_c G_p}{1 + G_c G_p}$$

which is the transfer function for the closed loop.

Evaluating the Closed Loop Transfer Function

The frequency response gain and phase lag for the closed loop might be obtained through the rather onerous application of mathematics, but Peter Harriott, in his excellent text (*Process Control*, McGraw-Hill, New York City, 1964), suggests an easier approach.

Since any component represented by a transfer function has a gain and a phase lag at every frequency, these two characteristics can be represented by a vector, and the overall gain and phase lag can then be determined by adding vectors to produce the vector for the closed loop.

To simplify matters, let the product G_cG_p be replaced by G, so that the closed loop transfer function becomes

$$\frac{G}{1+G}.$$

In Figure 9-4, the vector G has a length equivalent to the open loop gain, and it is orientated at an angle (a) equal to the open loop phase lag, at the selected frequency.

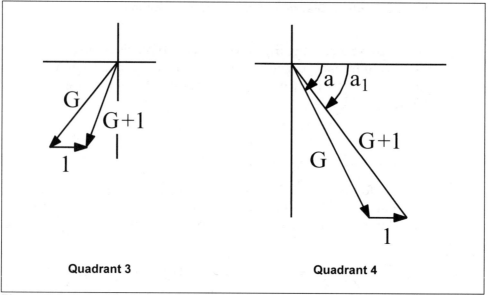

Figure 9-4. Using Vectors to Determine the Closed Loop Gain and Phase Angle.

The 1 vector, as a constant, has unit magnitude but no phase angle. Its direction is parallel to the horizontal axis and in the positive direction of the horizontal axis. The result of adding the G vector and the 1 vector diagrammatically is the vector representing G + 1.

At a somewhat higher frequency than that represented by the quadrant 4 diagram, the greater phase lag shifts the vector diagram into quadrant 3. The gain will likely be lower also, as indicated by the shorter G vector.

To find the closed loop gain and phase lag, it is necessary to divide the G vector by the G + 1 vector, since the form of the transfer function for the closed loop is a quotient. The procedure for dividing vectors of this type calls for *dividing* the magnitude of the vector in the numerator by the magnitude of the vector in the denominator, and *subtracting* the phase angle of the vector in the denominator from the phase angle of the vector in the numerator. Following this procedure yields the closed loop gain and phase lag at the particular frequency at which the open loop gain and phase lag were determined.

Specifically, this means that the closed loop gain will be equal to

$$\frac{\text{Magnitude of the G vector}}{\text{Magnitude of the }(G+1)\text{ vector}}$$

and the closed loop phase lag will be equal to $(a - a_1)$.

Figure 9-5 appears confusing but it contains an important fact. It represents the case in which the open loop phase lag is −180°. At this particular frequency, it is quite possible that the open loop gain will have fallen off to a value less than 1, which is why the G vector in Figure 9-5 is shown considerably shorter than in Figure 9-4. At a phase angle of −180°, all three vectors are going to lie on the horizontal axis.

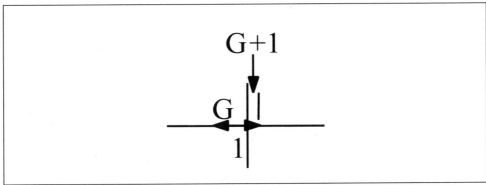

Figure 9-5. Both Vectors Have the Same Phase Lag at 180°.

The G vector, which represent the open loop, will point in the negative direction, in accordance with the −180° phase lag. If the open loop gain is less than 1, then when the unit vector is added, the resultant G + 1 vector will end on the positive side of the vertical axis, pointing in the positive direction. This indicates a phase angle of zero. Thus, when the phase angle of the G + 1 vector is subtracted from the open loop phase angle, the closed loop phase angle turns out to be −180° − 0°, or −180°, the same as the open loop phase angle.

The all important point here is that the frequency that creates a phase lag of −180° in the open loop also creates a −180° phase lag in the closed loop, and *this is the crucial frequency at which the closed loop is going to oscillate.* No other frequency will cause an identical phase lag in both the open and closed loops.

10

The Z–N Approximation

In Chapter 8, which deals with frequency response analysis, the point was made that attempting to obtain frequency response data on actual operating equipment was impractical. Process operators would not allow their equipment to be upset with sine wave disturbances over a lengthy time period, when their job was to ensure production and safe operations.

Nonetheless, the dynamic behavior of most operating equipment cannot be predicted by sitting in front of one's computer and applying theory. Real life operating equipment often consists of multiple time constants, with some dead time thrown in. The only way to get at the dynamic characteristics is with a field test. The question then becomes: Can a field test be devised that would be acceptable to plant operating people and would also yield the required information?

J.G. Ziegler, during his career as a control systems engineer with the Taylor Instrument Companies, answered this question and made a significant contribution to control systems analysis. His method has come to be called the *Ziegler–Nichols approximation*, or for short, the Z–N approximation.

Historical

The events that led up to the Z–N approximation are interesting and deserve some space. During the late 1930s and early 1940s, there was no ISA—The Instrumentation, Systems, and Automation Society. In the United States, any technical work on control systems that was felt to be noteworthy was made public through accredited professional societies such as the American Society of Mechanical Engineers or the American Institute of Chemical Engineers, at their national or regional conferences or through their publications. Some work on control theory was being done independently in the United States and England by individuals working in universities or manufacturing facilities, but the only organized group that had been officially formed up until that time was a Control Systems group, made up of several ASME members. As the one formal group that was recognized as such, the group members felt that they were in the forefront in the understanding of control systems theory.

To broaden their studies, the ASME group felt they needed an analogous system that exhibited typical process control characteristics. An obvious feature of feedback control systems is that they oscillate, and on this basis, the ASME group selected a mass oscillating at the end of a spring, with damping added to cause the oscillations to eventually die out. Thus, any conclusions that the ASME group reached with regard to control systems behavior were based on their analysis of their damped spring mass system.

It was at that time that Ziegler came to the Taylor Instruments headquarters at Rochester, N.Y., hoping to benefit from the knowledge of the resident control systems experts there. Much to his disappointment, he found that control theory, in their minds, was apparently limited to "capacity is good, and lags are bad."

Since this hardly satisfied his inquiring mind, Ziegler spent much of his time in the Taylor research lab working with a rudimentary process simulator. His technique consisted for the most part of sending a step change input into the simulator and watching what happened at its output. He collected numerous chart records of simulator outputs, and to these he gave the name *process reaction curves*. For our purposes, the term process reaction curve will mean the output response of a process, component, or whatever, to a *step change* in input.

One thing Ziegler noticed almost immediately was when the dynamics in the simulator were any more complex than those of a single time constant, the process reaction curve assumed an S shape. This happened because the output response did not start out from t = 0 at its maximum rate of change, as it would do if there were only a single time constant involved. The S shape showed that the response would begin slowly and then would build up to its maximum rate of change only after a period of time. Therefore, if a step change input is applied to any process or control system component, and the reaction curve that results shows the S shape, then the presence of something more than a single time constant, possibly multiple time constants or a time constant plus dead time, is indicated.

Figure 10-1A is a process reaction curve made by a single time constant. In this case, the output starts out from zero at its maximum rate of change.

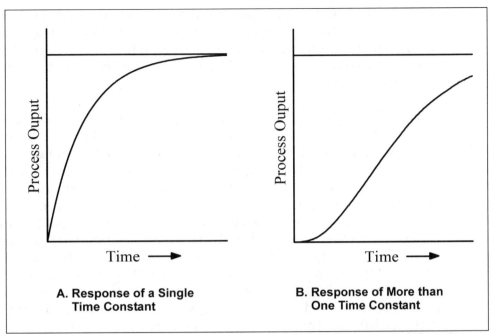

Figure 10-1. Response of a Single Time Constant (A) and Response of More than One Time Constant (B).

Figure 10-1B is a reaction curve made by two time constants in series. The reaction curve now starts out relatively slowly and builds up to its maximum rate of change later, thus creating the S shape.

Describing mathematically the reaction curve created by a single time constant is not a problem. This was done in Chapter 6, which dealt with differential equations. Describing the S shaped reaction curve in mathematical terms, however, requires a knowledge of mathematics, which most engineering graduates do not possess.

One person who did have this capability was N.B. Nichols, the Director of Research at Taylor Instruments while Ziegler was working in the research lab. Ziegler had a great admiration for Nichols' knowledge of mathematics and considered himself fortunate that he could call on Nichols when he needed help with the mathematics of control systems. In fact, Ziegler felt that the assistance he received from Nichols was valuable enough that he included Nichols' name as a co-author on some of the technical articles that Ziegler himself wrote.

Real life processes, when subjected to a step change disturbance, will inevitably exhibit the S shaped reaction curve. Since analysis using mathematics was out of the question, Ziegler was led to investigate the possibility of using a graphical method. As part of his ongoing research, Ziegler not only tested simulated processes by running process reaction curves, he also connected the processes to an automatic controller in an attempt to understand the pattern of controller settings which were required to have the simulated control system recover from a disturbance with the desired 1/4 amplitude ratio.

One point that particularly interested Ziegler was the controller gain setting which was required for optimum control, and whether anything could be learned from the process reaction curve that would give him a clue to the correct controller gain value. As his work progressed, he became convinced that two properties of the reaction curve had a significant effect on the correct gain setting. The first was the time interval at the start of the reaction curve during which the value of the curve hardly changed from the value at which it started out. The second was the maximum rate of change that the reaction curve eventually attained before it began to taper off to a new steady state value. To put numbers on these two properties, Ziegler drew a tangent to the reaction curve at the point of its steepest slope and extended the tangent downward until it reached the vertical t = 0 axis. This is illustrated in Figure 10-2.

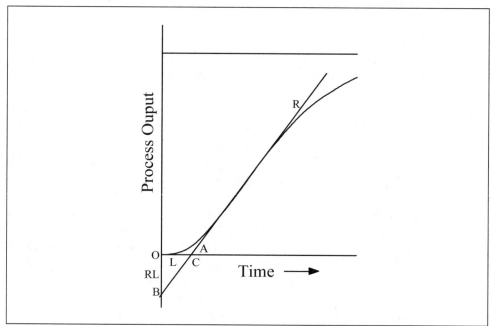

Figure 10-2. A Process Reaction Curve.

In the diagram, R is the slope of the tangent. Its units will be units of the process output per minute. The tangent intersects the horizontal line that represents the initial (t = 0) value of the process output at point C. The distance OC along this line is a time quantity and is designated L minutes.

The tangent then extends downward to intersect the vertical axis at point B. In this way, the mathematically complicated S curve has been replaced by the line segment OC, which is L time units in length, and the tangent with slope R.

If the controller gain setting depends on both R and L, then it is a reasonable bet that it depends on the product of them both, or RL. Since R is the slope of the tangent, then R is equal to the tangent of the angle A and also to the tangent of the angle OCB, which is equal to angle A. The tangent of angle OCB is equal to

$$\frac{OB}{OC} = \frac{OB}{L} = R, \text{ from which OB must be equal to RL.}$$

Ziegler's tests showed that if the quantity L, which is essentially dead time, increased, then a lower controller gain was required. The same applied if the slope R became greater. Could it be possible, therefore, that the required controller gain setting for a recovery with optimum stability depended on 1/RL? Numerous test runs that Ziegler conducted after developing this theory, proved that he was on the right track.

During the course of his work, Ziegler had occasion to talk informally with other engineers who were interested in what he was doing, and at one point the suggestion was made that he should prepare a paper for presentation at the ASME national conference in 1941. While this was encouraging, Ziegler knew that a paper that dealt with gain settings alone would not satisfy the audience. The following comment would obviously be made: That is all well and good, but how do we set the other controller settings, automatic reset, and derivative?

Further investigation by Ziegler, along with frequent consultations with Nichols, led him to the conclusion that the optimum reset and derivative settings had to be dependent on the apparent dead time L alone. As a result of some work by Nichols, the optimum values for reset and derivative were arrived at as simple functions of L. Thus, all of the ingredients for the complete paper were present.

The famous paper, *Optimum Settings for Automatic Controllers* (American Society of Mechanical Engineers, 1941) was then put together. Ziegler included their formulas for determining the optimum gain, reset, and derivative settings, based on first obtaining a process reaction curve. He also included a second approach, which involved finding the period of oscillation of the control system in the automatic mode, by turning up the gain of the controller until, in response to a disturbance, the system oscillated continuously on its own. The paper was then submitted.

To ensure that some questions would be asked in the question period following the presentation of a paper, it was customary to send out a copy of each paper to selected individuals prior to the conference. It was at this point that the storm broke. When the ASME Control Systems group received their preprint of Ziegler's paper, they hit the roof. What shook them was Ziegler's assertion that a process control system could be made to oscillate continuously. On the basis of their observations of their spring mass model, they claimed continuous oscillations were impossible.

As far as their model was concerned, they were right. A spring mass system will act the same as a pure two time constant system. Figure 10-3 shows the frequency response data for a system that contains two time constants and nothing more. The time constants are 2 minutes and 1 minute. The diagram shows that the phase lag curve, even with a second time constant in place, is not going to reach –180°, except at a theoretical infinite frequency. Therefore, a two time constant system cannot be made to oscillate on its own. The result was that the Control Systems group assailed the conference program committee and demanded that Ziegler's paper be withdrawn.

The fact that eluded these poor souls was that their model did not accurately represent a real life process control system. There will always be a time constant in the control valve, one or more time constants in the process, and at least one time constant in the measurement sensor, without counting any dead time that may have crept in as well. Any dynamic delay, no matter how small, that is present over and above the two time constants, will push the phase lag curve down below –180°, and make it possible for the control system to oscillate continuously on its own, given the correct amount of gain.

In the end, over the objections of the ASME Control Systems group, Ziegler was allowed to present the paper. Today, *Optimum Settings for Automatic Controllers* by Ziegler and Nichols would likely get most people's vote as the all time most significant article ever prepared on the subject of process control. Ziegler and Nichols both died in 1997, knowing that their article had received world wide acclaim.

The Z–N Approximation

Since most real life process reaction curves have the S shape and are complicated mathematically as a result, a practical question would be: Could it be possible to replace the S shaped curve with another reaction curve, which had dynamic properties equivalent to those of the S curve, but which was manageable mathematically through the use of transfer functions?

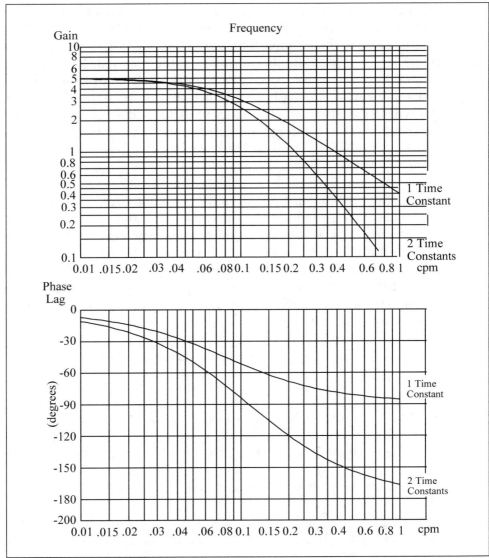

Figure 10-3. Frequency Response Data for Single and Two Time Constant Systems.

Ziegler's response to this question was the proposition that a synthetic process reaction curve that consisted of a dead time interval followed by a single time constant could be substituted for the S curve, thus rendering it possible to analyze the behavior of the complete control system. The frequency response data for dead time plus a single time constant in tandem can be easily derived. Unfortunately, in his paper, *Optimum Settings for Automatic Controllers*, Ziegler did not propose a method for determining the actual values that should be assigned to the dead time and time constant. Since we can no longer call upon Ziegler for assistance, we have to improvise.

Whatever method is contrived, the prime requirement will be that the replacement curve must generate a phase lag of −180° at a frequency that is at least close to the frequency at which the original S curve produces a phase lag of −180°. With this stipulation, the plan will be to start with a model process reaction curve that is sufficiently complex to generate an S shaped reaction curve, but for which the components are accurately known, so that the overall frequency response data can be accurately determined. Then a second curve consisting of dead time and a single time constant can be fitted to it and tested to see if it meets the important criterion.

The model process reaction curve will be one created by three time constants in series. The time constant values are 2.0, 1.5, and 0.5 minutes, respectively. Figure 10-4 shows the S shaped process reaction curve created by this combination. The phase lag curve for this synthetic process will cross the −180° line at a frequency of 0.26 cpm. The verification for this is not difficult. In general, the phase lag contributed by a time constant is equal to the angle whose tangent is −ωT, where ω is the frequency in radians per minute, and T is the time constant in minutes.

$$\omega = 2\,\pi \times f = 2\,\pi \times 0.26 = 1.634 \text{ rad}/\text{min}$$

For time constant 1, phase lag = $\tan^{-1}(-1.634 \times 2.0) = \tan^{-1}(-3.27) = -73.0°$

For time constant 2, phase lag = $\tan^{-1}(-1.634 \times 1.5) = \tan^{-1}(-2.45) = -67.8°$

For time constant 3, phase lag = $\tan^{-1}(-1.634 \times 0.5) = \tan^{-1}(-0.82) = -39.2°$

Total phase lag from all three time constants = −180.0°.

In Figure 10-4, the scale for the process variable has been selected arbitrarily on a 0 to 4 basis to simplify the analysis. It could be in temperature, pressure, or other process units, with an appropriate factor. The process reaction curve for the three time constant system starts out at 0 time and 0 on the vertical scale, proceeds through its S shape, and eventually reaches its ultimate value of 4 after a long time.

The line AB is the tangent to the reaction curve at its steepest point on the curve. The tangent meets the horizontal (time) axis at 0.85 minutes. This is the time value of the quantity L, in this case.

At first glance, it would appear that the search for the values for the dead time and the time constant of the replacement reaction curve has been successful. The dead time should be 0.85 minutes. Therefore, it only remains to find the value for a time constant that, in combination with the 0.85

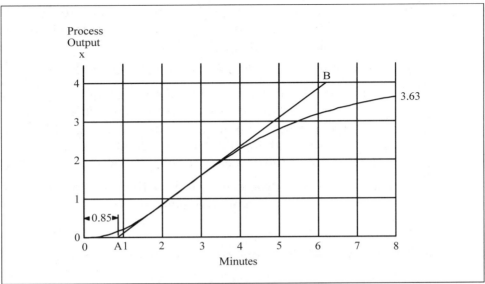

Figure 10-4. A Process Reaction Curve Created by Three Time Constants in Series.

minute dead time, will produce a phase lag of –180° at a frequency of 0.26 cpm.

Alas, it won't be that easy. How much phase lag does 0.85 minutes of dead time create at a frequency of 0.26 cpm?

$$0.26 \text{ cpm} = 1.634 \text{ rad/min, as before.}$$

$$\text{Phase lag} = -\omega L = -1.634 \times 0.85 = -1.39 \text{ rad} = -79.6°$$

The graph in Figure 10-3 shows that the maximum phase lag that one time constant can produce is –90°. Therefore, no single time constant, in combination with a dead time of 0.85 minutes, can produce –180° of phase lag at 0.26 cpm. L = 0.85 minute does not work.

There is another problem. It is known that the three time constant process will oscillate at a frequency of 0.26 cpm because it is known that the time constants are 2.0, 1.5, and 0.5 minutes. In the case of a real life process reaction curve, the dynamic elements involved are not known, so the frequency at which the control system will oscillate can't be determined by that route. Conclusion? Back to the drafting board.

The starting point in the search for the appropriate time constant is the three things that are known about it.

- In Figure 10-4, the time constant would start out from a point on the horizontal (time) axis, proceed upward in standard time constant fashion, and eventually reach a final value of x = 4.

- The point from which it starts must be a time value >0.85 minutes.

- The process variable y should track the path of the original three time constant reaction curve as closely as possible. One way that this could be achieved would be to have the single time constant curve intersect with the three time constant curve at its end point, at which t = 8, and the process variable x = 3.63.

- The values of the dead time L and the time constant T must create a phase lag of −180° at some frequency that is reasonably close to 0.26 cpm.

The function f(t), which is consistent with these conditions, is

$$x = 4\left(1 - e^{-\frac{1}{T}(t-L)}\right)..$$

The math will be easier to handle if $1/T$ is made equal to z. Then

$$x = 4\left(1 - e^{-z(t-L)}\right).$$

If t = 8 and x = 3.63 satisfy this expression, then $3.63 = 4(1 - e^{-z(8-L)})$.

Reducing this,

$$\frac{3.63}{4} = 1 - e^{-z(8-L)}, \text{ and } e^{-z(8-L)} = 1 - \frac{3.63}{4} = 0.0925$$

$$-z(8 - L) = \log_e 0.0925 = -2.38$$

$$z = \frac{2.38}{8-L}, \text{ and } T = \frac{1}{z} = \frac{8-L}{2.38}.$$

By using this relationship, it is possible to enter trial values for the dead time L and calculate the associated value of the time constant T, so that the time constant graph will begin at t = L and pass through the point t = 8, x = 3.63. This does not, however, yield the rest of the information that is really required, namely, the phase lags, which will be contributed by the dead time and the time constant, and their sum. The phase lag calculation cannot be done without knowing the frequency.

Estimating the Frequency of Oscillation

Optimum Settings for Automatic Controllers contains a clue as to how the frequency of oscillation can be estimated. Ziegler and Nichols concluded that the optimum controller reset rate and derivative settings depended on the dead time L alone. Accordingly, they included in the article the formulas for calculating the optimum settings for both procedures: finding the period of oscillation (P_u) or making a process reaction curve and finding the dead time L by drawing the tangent to the curve. The formulas they published were:

$$\text{Optimum derivative} = \frac{P_u}{8} = 0.5\,L$$

$$\text{Optimum reset rate} = \frac{2.0}{P_u} = \frac{0.5}{L}.$$

The constants that Ziegler and Nichols used in these relationships were based to some extent on the pneumatic controllers that Taylor Instruments were marketing in the 1940s, and on the internal pneumatic circuitry that these controllers used. Consequently, these formulas will not necessarily give the best values for the reset rate and derivative settings for the analog and digital electronic controllers in use today. However, the point here is not to be able to calculate reset and derivative settings. Rearranging either of these formulas, it is apparent that the period of oscillation P_u is equal to four times the dead time L and this does not depend on hardware. This is the information that is needed to make it possible to estimate the frequency of oscillation from any process reaction curve.

The tangent to the steepest point on the three time constant reaction curve meets the horizontal axis at 0.85 min. The period of oscillation is therefore estimated to be four times this, or 3.4 minutes, and the frequency of oscillation will be the inverse of the period, or 0.294 cpm. The true value is 0.26 cpm, so the value obtained by the estimation procedure is reasonably close. In the absence of any better method, this procedure will have to suffice.

Values for the Dead Time and Time Constant

Completing the example will best illustrate the method.

Starting with the relation

$$T = \frac{8 - L}{2.38}$$

select a trial value for L.

A previous calculation showed that L = 0.85 is too small a value, so the first trial value will be L = 1.0.

If $L = 1.0$, then $T = \dfrac{8 - 1}{2.38} = 2.94 \, \text{min} \cdot$

Frequency of oscillation is estimated to be 0.294 cpm = 1.847 rad/min.

Phase lag for L = 1.0 min. is $- 1.847 \times 1 = -1.847$ rad. $= -105.8°$.

Phase lag for T = 2.94 min. is $\tan^{-1} (-1.847 \times 2.94) = \tan^{-1} (-5.43) = -79.6°$.

Total phase lag $= -105.8 + (-79.6) = -185.4°$, which is 5.4° too much.

At this point it is apparent that it would be easy to design a computer program that would quickly calculate the values for the time constant T, the phase lags for the dead time and for the time constant, and the total phase lag, from trial values of L. By using such a program, the values sought turn out to be:

Dead Time L = 0.95 min. Time Constant T = 2.96 min.

Phase lag for L = 0.95 will be $-1.847 \times 0.95 = -1.755$ rad. $= -100.5°$.

Phase lag for T = 2.94 will be $\tan^{-1} (-1.847 \times 2.94) = \tan^{-1} (-5.47) = -79.6°$.

Total phase lag $= -100.5 + (-79.6) = -180.2°$.

Just How Good Is the Approximation?

It is now appropriate to investigate how well the dead time plus one time constant approximation matches the original three time constant process reaction curve. This should be done in two ways:

- By overlaying the dead time plus one time constant reaction curve on the three time constant reaction curve (Figure 10-4).

- By comparing the frequency response gain and phase data for both reaction curves.

Figure 10-5 shows how the two reaction curves compare on a time basis. As planned, the curves intersect at t = 8, x = 3.63, although the dead time (L) values are 0.85 and 0.95 minutes, respectively.

What is most significant, however, is the comparison of the two phase lag curves. Figure 10-6 shows that they track fairly well, including at the point where they reach –180°. The original process reaction curve consisting of

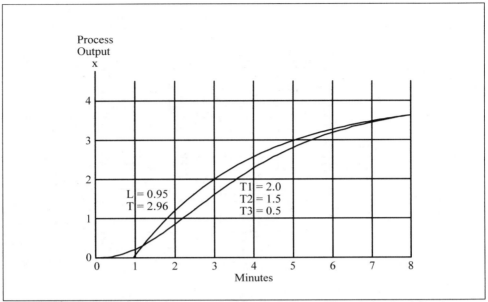

Figure 10-5. Process Reaction Curves for the Three Time Constant System and the Z-N Approximation.

three time constants shows a frequency of oscillation of 0.26 cpm, while the approximation reaction curve, with dead time and a single time constant, crosses –180° at a frequency of 0.29 cpm. The equivalent periods of oscillation are therefore 3.8 and 3.4 minutes. This is about the level of accuracy that is to be expected from the approximation.

At the frequency of oscillation, the gain values are considerably different, of the order of 0.3 and 0.75. This is not a problem, however, because the adjustable gain in the automatic controller makes is possible to set up the overall control system gain, which includes the gain of the process, to the value that produces a recovery with a 1/4 amplitude ratio.

Making a Process Reaction Curve

The process reaction curve was given its name in Ziegler and Nichols original 1941 article. Constructing a process reaction curve is really a dynamic test in which the input test signal is a step change. A test using this particular input pattern is a very useful one, in that it is easy to generate and it yields most, if not all, of the dynamic data needed to make a practical analysis of a process control system.

A process reaction curve is made with the automatic controller on Manual control. The procedure:

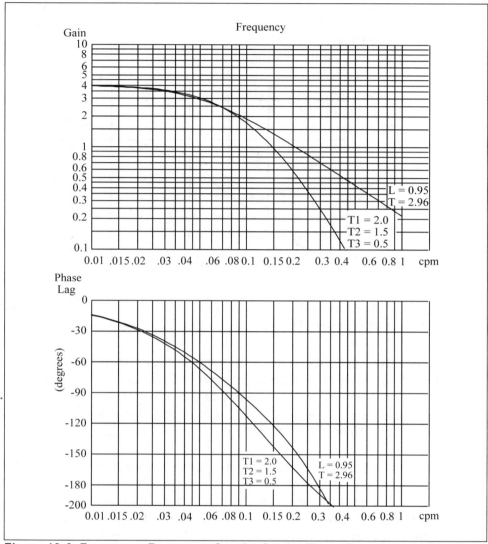

Figure 10-6. Frequency Response Graphs for the Three Time Constant System and the Z–N Approximation.

1.   Set the controller on Manual and allow the control system and the controlled variable to stabilize. Record the manual output to the control valve.

2.   Step the manual output to the control valve quickly to a new value. The size of the step change must be great enough that the controlled variable, when plotted either manually or automatically, exhibits the familiar S shaped curve, as shown in Figure 10-7. If no better information is available, then a 5% change in output to the valve is a reasonable number to start with. Record the size of the step change, Δm, in percent output signal to the control valve.

Once again, bear in mind that this test is being done on an operating plant. Consequently, it can be done only with the concurrence of the plant operator. One usually finds that if the operator can be convinced that running the test will ultimately lead to better control, he or she will usually agree to the test. On most processes the equivalent data will be obtained whether the step change drives the controlled variable up or down. Ask the operator which he or she would prefer.

3.   The test is begun when the step change in output to the control valve is made. Starting from this moment, record corresponding readings of time and the controlled variable; in Figure 10-7, it is the process temperature. Continue taking readings until it is obvious that the rate of change of the process variable is approaching its new steady state value. The readings should be spaced closely enough that they can be plotted later to produce a reasonably representative process reaction curve.

4.   When Step 3 is completed, set the output signal to the control valve, still on manual control, back to the original value recorded in Step 1. If this procedure is followed, the only disturbance that the process will see is a hump, upward or downward, in the controlled variable, as shown in Figure 10-7.

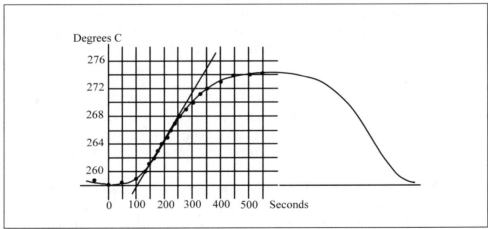

Figure 10-7. A Process Reaction Curve Test from an Actual Process.

Before describing what should be done with the data obtained, the following points on technique for taking the data should be noted. Ordinarily, a team of two observers will be required.

The documentation is made easier if the data sheet is set up beforehand. This is where another problem can be inadvertently created. It would seem that a logical approach would be to concentrate on the independent variable, that is, the time values. When the data sheet is laid out, the times, usually at about 15 second intervals, are written in. Then when the test begins, it is up to the team member who is tracking the process variable to call out its values as his partner calls out the times.

We now have the potential for a real confrontation. If the process variable is displayed on an analog indicator or recorder, and the value is in between two of the divisions, it may be difficult to decide quickly enough what the value of the controlled variable really is. In fact, it may happen that before the team member who has this job has figured out the value at one time point, his partner is calling for the next value. At this stage, there may be some pointed criticism about the capabilities of the harassed team member, with the result that he quits and goes back to the office or shop.

This problem should not occur if it is recognized that the time device, stop watch or otherwise, is the easier instrument to read. Some timers even have a digital readout. Consequently, a better procedure is to set down even division and mid division values of the controlled variable, and as the test proceeds, record the time values at which the controlled variable attains these values.

The ultimate arrangement, of course, would be to have a high speed recorder to record the values of the controlled variable on a time base over the course of the process reaction curve run. Unfortunately, not many engineering departments will have this equipment. However,one person with a cassette recorder might be able to do the job adequately.

The measuring system of an automatic controller contributes its share to the overall dynamics of the control loop. Consequently, the readings of the process variable which are taken must come from the measuring system of the automatic controller, not from some other meter which may be measuring the same process variable, and which may be easier to read.

At this point, a word of caution is appropriate. The procedure just described works only for processes that exhibit self regulation. Self regulation means that for each position of the control valve, as set on manual control, the controlled process variable will settle out at some reasonable value. Unfortunately, there are processes which are not self regulating. If a process lacking in self regulation is holding steady on manual control at $t = 0$, an unbalance which is subsequently caused by a step movement of the control valve will be integrated over time. The controlled variable will not settle at some new steady state value, but will drift up to maximum or down to zero.

The good news, however, is that most of the non self regulating processes have reasonably good dynamic behavior, are easier to control, and thus have no real need for obtaining dynamic data by making a process reaction curve, or by any other test. Process reaction curves are generally reserved for the hard jobs.

Example 1: Reaction of a Real Process

Inasmuch as the process reaction curve shown in Figure 10-7 is from an actual process, it would be meaningful to determine the values of the dead time and the time constant, and complete the approximation. Figure 10-8 shows the same process reaction curve with the trailing end, after the cut-off point, deleted. The temperature scale has been replaced with an arbitrary scale that has its zero point at the level from which the reaction curve starts out. Since the two values being sought—the dead time and the time constant—are both time values, changing the scale of the dependent variable will not affect the outcome.

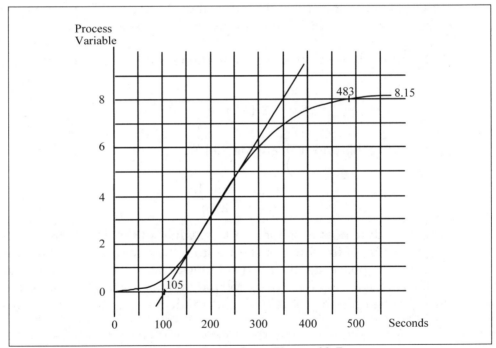

Figure 10-8. Process Reaction Curve from Figure 10-7.

The equation for the time constant, as before, is,

$$PV\ (Process\ Variable) = X \left[1 - e^{-\frac{1}{T}(t-L)} \right].$$

In this equation, X is the difference between the start of test and end of test steady state values of the controlled variable, T is the time constant, and L is the dead time.

The curve appears to show that the ultimate value X is 8.15 on the PV scale. The tangent to the curve at its steepest point crosses the base line at about 105 seconds. Thus, the estimated period of oscillation will be four times this value or 420 seconds. The frequency of oscillation will be 1/420 Hz, and in radians per second, the frequency will be

$$\omega \ = \ 2\pi \times \frac{1}{420} \ = \ 0.0150 \text{ rad/s}.$$

A point near the end of the reaction curve is also needed. A logical question would be: Why not use the point at the end of the curve where the PV is 8.15 (its ultimate value) and t = 550 s? This does not compute, however, because an inherent characteristic of any time constant is that it does not attain its ultimate value until t equals infinity. Therefore, some other point, near, but not at, the end of the curve, must be chosen. The curve appears to cross PV = 8 at t = 483 s, so this point will be used.

Substituting these values in the equation for the time constant gives,

$$8.0 \ = \ 8.15 \left[1 - e^{-\frac{1}{T}(483 - L)} \right].$$

Simplifying this and rearranging yields,

$$T \ = \ \frac{483 - L}{4.00}.$$

The procedure now calls for selecting values of the dead time L, determining the value of the time constant T from the relation above, and then calculating the phase lag that the dead time and time constant in combination will create at a frequency of 0.0150 rad/s. This trial and error routine is to be repeated until the values of L and T that create a phase lag of −180° have been found.

For this process reaction curve the values turn out to be L = 150 s and T = 83.3 s. The phase lag values are:

Dead time phase lag = $- \omega L = - 0.0150 \times 150 = -2.24$ rad = $-128.6°$.

Time constant phase lag = $\tan^{-1}(- \omega T) = \tan^{-1}(- 0.0150 \times 83.3) = - 0.895$ rad = $- 51.3$ deg.

Total phase lag = – 128.6 + (– 51.3) = – 179.9°.

Figure 10-9 shows how closely the Z-N approximation matches the original process reaction curve.

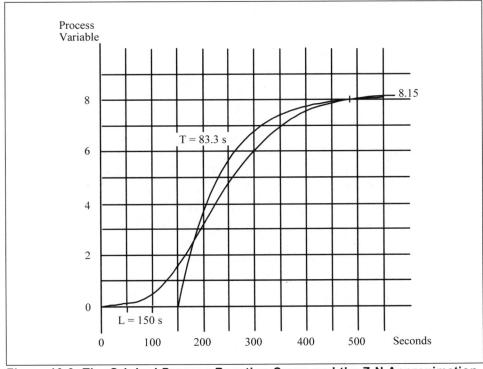

Figure 10-9. The Original Process Reaction Curve and the Z-N Approximation.

11

Units, Best Values, Formulas, and Other Good Stuff

True Value

Statements about the accuracy of numbers imply that an accurate, or true value, is known. Measurements that are based essentially on counts, where the items counted are people, money, pulses in digital systems, and so on, can be considered accurate. A statement that a new computer costs $1499 means exactly 1499 dollars.

On the other hand, for measurements of analog variables, and this includes nearly all process measurements, such as temperature and pressure, the exact value is known only to the Great Creator. The value ultimately obtained will only be the best approximation that we humans can make. Even if we are able to get the first 6 digits correct, it is still an approximation. A good quality temperature sensor might output a reading of 68.5°C, but this does not mean that the temperature is exactly 68.500 000 000... It simply means that the true temperature lies somewhere between 68.45° and 68.55°. Still, up until about 1960, engineering calculations were done on sliderules, which were capable of producing only 3 significant figures, but this was good enough to design oil refineries that would produce products and airplanes that would fly.

In industry, the true value of any process measurement is usually taken to be the value indicated by some standard measurement device. Unfortunately, the majority of measurement standards are not rugged enough to tolerate the plant environment and have to be kept in a laboratory. The sensors that are used in the process operations are then calibrated against the standard.

For this procedure to give satisfactory results, the accuracy of the standard must necessarily be two or three orders better than that of the plant sensor. A fact that helps the situation is that process measurements nearly always have a tolerance that is acceptable for the successful production of products. A temperature measurement that is within ±1 degree of the true value might well be sufficiently accurate for operating a plant.

Errors

It is customary to express the accuracy of a measurement in terms of the error that can be expected. The error is defined as the difference between the observed value of the measurement (OV) and the true or accurate value (TV). Thus, if the error is designated ε, then

$$\varepsilon = OV - TV.$$

An error can be expressed in actual units of measurement or as a fraction or percentage of the true value. An error can be positive or negative, depending on whether the observed value is greater or less than the true value.

Sometimes statements about the accuracy of measurements tend to get sloppy. Such a statement might be "Orifice meters are 3% accurate." This statement could be interpreted incorrectly in either of two ways. First, the statement that the meter is 3% accurate implies that it could be 97% in error, which is unrealistic. Second is the implication that no matter what the reading of the meter might be, it will be in error by 3%. This is not realistic either because most industrial meters can usually be calibrated so that they are accurate at at least two points in their range of measurement. A proper statement of accuracy for the meter is "This meter is accurate to within 3% (of the true value)." The key word is "within."

Errors in Combinations of Quantities

Two quantities, whose true values are X and Y, are known to have possible errors of ε_x and ε_y. If the two quantities are to be added, what will be the error in the sum? If X and Y are the true values, then the measured values will be $X + \varepsilon_x$ and $Y + \varepsilon_y$, bearing in mind that the errors could be positive or negative.

$$(X + \varepsilon_x) + (Y + \varepsilon_y) = (X + Y) + (\varepsilon + \varepsilon_y)$$

The conclusion drawn from this is that *when quantities with known errors are added or subtracted, the **actual** errors are added*. Note that errors are always added, never subtracted, even if the quantities to which the errors belong are subtracted.

If the two quantities X and Y are to be multiplied, then

$$(X + \varepsilon_x) \times (Y + \varepsilon_y) = XY + Y\varepsilon_x + X\varepsilon_y + \varepsilon_x\varepsilon_y.$$

Since ε_x and ε_y are hopefully small compared with X and Y, for the purposes of this calculation, their product $\varepsilon_x\varepsilon_y$ can be ignored, Then,

$$XY \;+\; Y\varepsilon_x \;+\; X\varepsilon_y \;=\; XY \;+\; XY\left(\frac{\varepsilon_x}{X} \;+\; \frac{\varepsilon_y}{Y}\right).$$

The two factors within the brackets are the fractional errors of X and Y. Their sum is the fractional error in the product XY. The sum of the fractional errors multiplied by the product XY, as shown in the expression, will be the actual potential error in the product XY, in whatever units X and Y are measured.

The development for the quotient of X over Y produces the same result. Accordingly, the rule is: *The potential **fractional** error of the product or the quotient of quantities is the sum of the **fractional** errors of the individual quantities.*

Since percentage error is simply another form of fractional error, the words "percentage errors" could be substituted for fractional errors in the rule.

Correction Factor

A correction factor (CF) can be determined if the error is known. It is the quantity that should be added to the observed value in order to correct it to the true value. As a mathematical expression,

$$OV + CF = TV.$$

Rearranging this, $OV - TV = -CF.$

Recalling that the expression for the error was $\varepsilon = OV - TV$, then the correction factor CF must be equal to the *negative* of the error. Thus, if there is a scale that is showing a weight of 0.5 kg when it should be reading zero, then the correction factor for any reading taken from this scale should be –0.5 kg.

Significant Figures

Simply put, significant figures are numbers that actually mean what they say. It is definitely possible for numbers that have no meaning or fact to them to emerge as a result of a calculation. This is particularly true today when there are pocket calculators which will fill the whole viewing screen with numbers whether they have any meaning or not.

Suppose that the following population data is available for the City of Edmonton; the City of Ft. McMurray, 500 km to the north; and the village of Wandering River, which is halfway in between.

City of Edmonton	941,000
City of Ft. McMurray	43,900
Village of Wandering River	63

The total population of all three places calculates out to be 984,963 but obviously not all of the six figures have any real meaning.

The figure given for the City of Edmonton really specifies that the population is somewhere between 940 500 and 941 500. In other words, there will be a tolerance in the sum of ±500 persons. In view of this, the population figure for Wandering River has no meaning at all in the sum, while the hundreds digit for Ft. McMurray is relevant, in calculating the sum, only to the extent of bumping the thousands digit from 3 to 4.

When due regard is given to the real significance of the figures, the total population of the three communities should be recorded as 985 000.

The loss of meaning, or significance, always occurs in the trailing digits. The question then becomes: How many of the leading digits are significant? In the case of a sum or difference, *the last digit in the sum or difference that is significant, will be the same as the last significant digit in the greatest term in the summation.*

Until the 1960s, significant figures in products or quotients were not really a problem, since the calculations were done on sliderules, which could generate only three figure answers. Calculators that have come along since that time can create a false impression. Consider the operation of multiplying 102.7 by 3.14. If this is plugged into a calculator, the answer comes out 322.478, but how many of these figures are really significant?

If the two numbers are measurements of some kind (the decimal places suggest that they are), then the number 102.7 says that the measurement it represents lies between 102.65 and 102.75. Similarly the measurement that has a value of 3.14 lies somewhere between 3.135 and 3.145.

Multiplying the two lower values gives $102.65 \times 3.135 = 321.808$.

Multiplying the two higher values gives $102.75 \times 1.345 = 323.149$.

Thus, all that is known for sure about the product of the two numbers is that it lies between 321.808 and 323.149. Consequently, in the number cranked out by the calculator, 322.478, the numbers following the decimal point are meaningless. The only significant digits are 322.

The rule for the significant figures in a product or quotient is that *the number of significant figures in the product or quotient can be no greater than the number of significant figures in the term that has the least number of significant figures*. In the case of the example, the number 3.14 has the fewest significant figures (three), therefore no more than the first three figures in the calculated product will be significant.

Special attention should be given to the matter of trailing zeros after the decimal point. The fact that the relationship between inches and centimetres is written 1 in. = 2.54 cm does not imply that all of the figures after the 4 are zeros. However, it turns out that the next figure after the 4 is in fact a zero, so that the relationship should be recorded as 1 in. = 2.540 cm. This shows that the relationship is known to four significant figures, not three, and furthermore that the fourth significant figure is a zero.

Conversion of Units

Twice during the last few years, somewhat similar incidents were reported by the news media. An Air Canada jet liner ran out of fuel in mid air. Fortunately, the captain was one of the very few who had the necessary skill to land the plane on zero power. The problem was attributed to a mix up between pounds and kilograms of fuel.

Later, NASA crashed a space probe that was designed to land on Mars. The cost of the probe was some hundreds of millions of dollars, not counting the salaries of the highly trained technical people who had to track and guide it through its months long journey from Earth. The cause was said to be a mix up between feet and metres in the rate of descent.

Failure to convert correctly from one system of units to another can have serious consequences, enough so that some attention to a conversion procedure is justifiable. The process of converting measurements in one system of units into another system is sometimes called *scaling*. There are likely numerous methods of performing conversions. The one that will be described here, however, is straightforward and virtually foolproof. The explanation will be clearest if an example is worked.

Your neighbor has come to you with a problem. He has bought some fertilizer for his lawn The directions on the bag say to apply it at a rate of 2 kg per 100 m². He is in difficulty because he knows the area of his lawn is 2 560 ft². In addition, the only scale he has measures in pounds, not kilograms. How much fertilizer should he put on his lawn?

The first step is to convert 2 kg into pounds. Two statements are required, the first being any known *correct* relationship between pounds and kilograms. You happen to know that 1 lb = 0.453 6 kg, so you use this as the first statement. The second statement is the relationship you need to know, using an x, or a ? for the unknown quantity. The setup looks like this.

> 1st statement   1 lb = 0.453 6 kg
>
> 2nd statement   ? = 2 kg.

When these statements are written down as shown, it is *vital* to ensure that the same units in both statements are on the same side of the equals sign. In this case, pounds are on the left side of the equals sign, kilograms on the right, in both statements. It would make no difference if this were reversed (kilograms on the left, pounds on the right), as long as consistency between the two statements is maintained.

With the two statements in place, mentally draw two diagonals across the four numbers.

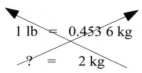

The unknown quantity, (?), will always be equal to the *product* of the two numbers that lie on the path that does *not* go through the (?), divided by the single number that is on the path that includes the (?). Therefore,

$$? = \frac{2 \times 1}{0.453\ 6} = 4.41 \text{ lb}.$$

Use the same procedure to convert 100 m² into ft². You recall that 1 ft = 0.304 8 m. Therefore, 1 ft² = 0.304 8² m² = 0.092 9 m².

> 1st statement   1 ft² = 0.092 9 m²
>
> 2nd statement   ? = 100 m²

$$\text{Therefore, } ? \ = \ \frac{100 \times 1}{0.092\ 9} \ = \ 1\ 076\ \text{ft}^2 .$$

The specified application rate of 2 kg per 100 m^2 is therefore equivalent to 4.41 lb per 1 076 ft^2. For 2 560 ft^2,

$$4.41\ \text{lb} = 1\ 076\ \text{ft}^2$$

$$? \ = 2\ 560\ \text{ft}^2$$

$$? \ = \ \frac{2\ 560 \times 4.41}{1\ 076} \ = \ 10.5\ \text{lb} .$$

When this method is used, it doesn't matter which units are on which side of the equals sign. In fact, it doesn't matter if the statement of the known relationship is the first statement or the second one. What does matter is that in both statements, consistency in the placement of the units with respect to the equals sign is maintained. If this is done then a correct conversion will result.

Converting Formulas to New Units

It sometimes happens that a formula is needed to calculate some required quantity. Unfortunately, the available formula uses a system of units that is not convenient. The problem is to convert the available formula to the units that one wishes to use.

Since a mistake in converting the formula's units will inevitably lead to calculating the wrong value for the required quantity, the procedure that is used should be logical and as error proof as possible. First of all, consider these two statements, both correct, which follow.

$$1\ \text{lb} = 0.453\ 6\ \text{kg} \quad \text{Therefore, lb} \times 0.453\ 6 = \text{kg}$$

It is the second of these two statements that is needed to convert a formula correctly. Specifically, conversion factors are needed that will convert the desired units into the units required by the formula. Correct conversion of the formula depends, therefore, on coming up with the right conversion factors.

The recommended procedure is to set up the conversion statements, similar to the two above, with the *desired* units on the *left side* of the equals sign, and the units *required* by the formula on the *right side* of the equals sign. The whole process will be illustrated best by working an example.

If the inside diameter of a pipe, the velocity of the flowing stream, and the density of the stream are known, then the mass flow rate (W) of the stream in the pipe will be

$$W \text{ kg/s} = \pi/4 \text{ v } d^2 \rho = 0.785 \text{ 4 v } d^2 \rho.$$

In this expression, v is the velocity in per second, d is the inside diameter of the pipe in inches, and ρ is the density of the stream in kilograms per cubic metre. What is required is the equivalent expression with the velocity in feet per second, the inside diameter in inches, and the density in pounds per cubic foot. The starting point is to set up the conversion statements, one at a time, with the desired units on the left and the required units on the right.

For the velocity v, 1 fps = 0.304 8 m/s. Therefore fps × 0.3048 = m/s.

For the diameter d, $1 \text{ in.} = \dfrac{1}{12} \times 0.304 \text{ 8m} = 0.025 \text{ 4m}$.

Therefore $\text{in}^2 \times 0.025 \text{ 4}^2 = \text{m}^2$.

For the density ρ, $1 \text{lb/ft}^3 = 0.453 \text{ 6 kg/ft}^3 = 0.453 \text{ 6} \times \dfrac{1}{0.304 \text{ 8}^3} \text{ kg/m}^3$

$$= 16.02 \text{ kg/m}^3.$$

Therefore, $\text{lb/ft}^3 \times 16.02 = \text{kg/m}^3$.

If v_1, d_1, and ρ_1 are the velocity in fps, the diameter in in., and the density in lb/ft^3, then the formula converts to

$$W = 0.785 \text{ 4 } (0.304 \text{ 8 } v_1) \, (0.025 \text{ 4}^2 \, d_1^2) \, (16.02 \, \rho_1)$$

$$= 0.002 \text{ 474 } v_1 \, d_1^2 \, \rho_1.$$

Obviously, this result needs to be tested with some actual numbers. Suppose that in a 6 in. I.D. pipe, a fluid with a density of 50 lb/ft^3 is flowing with a velocity of 5 fps.

$$5 \text{ fps} = 5 \times 0.304 \text{ 8 m/s} = 1.52 \text{ 4 m/s}$$

$$6 \text{ in.} = 6 \times \frac{1}{12} \times 0.304 \text{ 8 m} = 0.152 \text{ 4 m}$$

$$50 \text{ lb/ft}^3 = 50 \times 0.453\ 6 \times \frac{1}{0.304\ 8^3} \text{kg/m}^3 = 800.9 \text{ kg/m}^3$$

Applying the original formula using the metric units,

$$W = 0.785\ 4 \times 1.524 \times 0.152\ 4^2 \times 800.9 = 22.26 \text{ kg/s.}$$

If the formula has been correctly converted, it should produce the same number, using the fps units.

$$W = 0.002\ 474 \times 5 \times 6^2 \times 50 = 22.26 \text{ kg/s.}$$

While some satisfaction can be derived from this, it points out that the job of converting the formula is not yet finished. The new converted formula can accept the fps units and give the correct answer, but it still produces the answer in kg/s, whereas the result is required in pounds per hour. The final step, therefore, is to modify the numerical factor 0.002 474 so that lb/hr are the units of the result.

$$1 \text{kg/s} = \frac{1}{0.453\ 6} \text{ lb/s} = \frac{1}{0.453\ 6} \times 3\ 600 \text{ lb/hr} = 7\ 937 \text{ lb/hr}$$

Therefore the numerical factor for the formula should be

$$0.002\ 474 \times 7\ 937 = 19.63.$$

The final result is

$$W \text{ (lb/hr)} = 19.63 \text{ v d}^2 \text{ } \rho, \text{ with v in ft/s, d in in., and } \rho \text{ in lb/ft}^3.$$

Example 1: Reynolds Number

In the SI metric system, the basic relationship for computing the Reynolds number Re is

$$Re = \frac{v d \rho}{\mu}.$$

In this relationship,

 v = the velocity of the stream in metres per second (m/s)
 d = the internal diameter of the pipe in metres (m)
 ρ = the density of the fluid in kilograms per cubic metre (kg/m^3)
 μ = the absolute viscosity of the fluid in Pascal seconds (Pa.s)

When the SI metric relationship is used, the constant multiplying the expression is 1.00.

Starting with the basic metric relationship, the assignment is to develop a formula for the Reynolds number having the form

$$Re = C\frac{Q\,sg}{d\mu}.$$

where

    Q = the volume flow rate of the liquid in barrels per hour (bph)
    sg = the specific gravity of the liquid
    d = the internal diameter of the pipe in inches (in.)
    μ = the absolute viscosity of the liquid in centipoise (cps)
    C = the constant that is required to accommodate the new units (to be
        determined).

The first step is to convert the metric formula so that it is in terms of volume flow rate and specific gravity instead of velocity and density. Volume flow rate is the product of the stream velocity and the cross sectional area of the pipe. In appropriate units,

$$Q = \frac{\pi}{4}d^2\,v, \text{ from which } dv = \frac{4Q}{\pi d}.$$

The specific gravity $sg = \dfrac{\rho}{\rho_s}$

where ρ_s is the density of water at 15°C.

Thus, $\rho = sg \times 999.1 \text{ kg/m}^3$. Inserting these values in the original formula,

$$Re = \frac{4Q}{\pi d} \times 999.1sg \times \frac{1}{\mu} = 1\,272\frac{Q\,sg}{d\mu}, \text{ all in metric units.}$$

The next step is to convert the individual terms, bearing in mind that the procedure calls for keeping the desired units on the left side of the equals sign.

$$1\text{ bbl} = 159.0 \text{ dm}^3 = 0.159 \text{ m}^3$$

$$1\text{ bph} = 0.159 \times \frac{1}{3\,600}\text{m}^3/\text{s} = 0.000\,044\,17 \text{ m}^3/\text{s}$$

$$\therefore \text{ bph} \times 0.000\,044\,17 = \text{m}^3/\text{s}$$

$$1\text{ in.} = \frac{1}{12}\text{ft} = \frac{1}{12} \times 0.304\,8 \text{ m} = 0.025\,4 \text{ m}$$

$$\therefore \text{ in. } \times 0.025\ 4 = \text{m}$$

$$1 \text{ cps} = 1 \text{ mPa.s} = 0.001 \text{ Pa.s}$$

$$\therefore \text{ cps} \times 0.001 = \text{Pa.s}$$

Substituting these values into the metric relationship for Re,

$$\text{Re} = \frac{1\ 272 \times 0.000\ 044\ 17\ Q \times sg}{0.025\ 4\ d \times 0.001\mu} = 2\ 212\frac{Q\ sg}{d\mu}\text{, in the desired units.}$$

This result should be tested with some actual data. In the test application, a liquid stream with a specific gravity of 0.80 and a viscosity of 0.70 cps is flowing through a pipe with an internal diameter of 6 in. at a rate of 500 bph. The Reynolds number computes to be,

$$\text{Re} = \frac{2\ 212 \times 500 \times 0.80}{6.0 \times 0.70} = 211\ 000.$$

Verifying this value for Re,

Flow rate $Q = 500 \times 0.000\ 044\ 17 = 0.022\ 09 \text{ m}^3/\text{s}$

Density $\rho = 0.80 \times 999.1 = 799.3 \text{ kg/m}^3$

Viscosity $\mu = 0.70 \text{ cps} = 0.70 \text{ mPa.s} = 0.000\ 70 \text{ Pa.s}$

Diameter $d = 6 \times 0.0254 = 0.152\ 4 \text{ m}$

$$\text{Velocity} = \frac{Q}{\text{Area}} = \frac{0.022\ 09}{\frac{\pi}{4} \times 0.152\ 4^2} = 1.211 \text{ m/s}$$

Inserting these values in the original metric formula,

$$\text{Re} = \frac{1.211 \times 0.152\ 4 \times 799.3}{0.000\ 70} = 211\ 000.$$

Example 2: Pressure of a Column of Liquid

One situation that comes up frequently concerns the pressure built up by a head of liquid. When a textbook formula is found, it usually does not employ the most convenient units. The basic relationship in the SI system is

$$p = \rho\, g\, h.$$

The pressure is in Pascals when ρ is in kg/m^3, g is in m/s^2, and h is in m.

The formula desired is one that gives the pressure (p) in psi, with the density expressed as specific gravity (sg), and the head of liquid (h) in feet.

By definition,

$$\text{specific gravity} = \frac{\text{Density of the liquid } (kg/m^3)}{\text{Density of water at } 15^O \text{ C } (kg/m^3)} = \frac{\rho}{999.1}.$$

$$\therefore \rho = 999.1 \times sg$$

Also, in the SI system, g = 9.812 m/s², and p (kPa) = p (Pa) ÷ 1000.

Substituting these facts into the basic formula gives,

$$p\,(kPa) = \frac{(999.1 \times sg) \times 9.812 \times h}{1000}$$

$$= 9.803 \times sg \times h\,(m).$$

The formula is required to accept h in feet. Following the procedure:

$$1 \text{ ft} = 0.304\ 8 \text{ m.} \quad \therefore \text{ ft} \times 0.304\ 8 = m$$

$$9.803 \times sg \times h\,(m) = 9.803 \times sg \times 0.304\ 8 \times h\,(ft) = 2.988 \text{ sg h (ft).}$$

For the resulting pressure to be in psi, not kPa:

$$1 \text{ psi} = 6.895 \text{ kPa}$$

$$? = 2.988 \text{ sg h kPa}$$

$$? = \frac{1 \times 2.988 \text{ sg h}}{6.895} = 0.433 \text{ sg h psi.}$$

Finally, p (psi) = 0.433 sg h (ft).

The Most Representative Value (MRV)

Sometimes a number of measurements of the same entity are made, possibly at different times. This leads to the question: What one value could be chosen as being the most representative value for the entity of concern?

The daily price of gasoline, in cents per litre, over a 15 day period, was noted to be:

62.5	64.5	65.5	66.5	61.5
61.5	62.5	66.5	67.5	67.5
68.5	68.5	66.5	65.5	63.5

Which value would be most representative of the price of gasoline over the 15 day period? There are actually four possibilities.

Average Value

The average value, or what the statisticians call the *arithmetic mean*, is probably the most familiar. To calculate the average, the 15 readings are summed, and the sum is divided by the number of readings, i.e. 15. For this example, the average turns out to be 65.2¢/l.

Although the average value is often used as the most representative value, there is obviously a problem in this case because the value 65.2 does not appear anywhere in the list of readings. This raises some doubt as to its appropriateness as the representative value.

Weighted Average

When a number of readings are taken, it may happen that some readings will be more relevant than others. At the beginning of the year, one might be attempting to estimate how much of one's income will go into paying for natural gas. The cost of gas for each of the last five years is available.

A simple average may not be as accurate as a weighted average. It would be more practical to assume that the cost of gas in more recent years should be given greater influence than the cost in earlier years. If G5, G4, G3, G2, and G1 were the yearly natural gas costs for each of the last five years, C1 being the cost for the year just ended, then the weights might logically be assigned as follows.

$$G5 \times 1 \qquad G4 \times 2 \qquad G3 \times 3 \qquad G2 \times 4 \qquad G1 \times 5$$

In this weighting, the cost of gas for the year just ended would receive five times the emphasis as the cost five years ago. Giving one of the readings a weight of five is equivalent to putting that reading into the sum five times rather than once, and this fact must be taken into account in the denominator of the expression. Consequently, the mathematical expression for the weighted average will be:

$$\text{Weighted Average} = \frac{G5 \times 1 + G4 \times 2 + G3 \times 3 + G2 \times 4 + G1 \times 5}{1 + 2 + 3 + 4 + 5}$$

For the gasoline prices, it may be considered that the latest five readings should receive twice the weight of the first five, and the second group of five readings should have 1.5 times the weight of the first five readings.

The sum of the first five readings is 320.5.

The sum of the second five readings is 325.5.

The sum of the last five readings is 332.5.

The weighted average equals

$$\frac{320.5 \times 1 + 325.5 \times 1.5 + 332.5 \times 2}{5 \times 1 + 5 \times 1.5 + 5 \times 2} = \frac{1\ 473.75}{22.5} = 65.5\,.$$

The fact that the digit after the decimal place turned out to be a 5, the same as for the individual readings, is just a coincidence. If any of the 15 readings had had a different value, then the third digit in the weighted average would have been something other than a 5.

Mode

The *mode* is the reading that appears most often in the set of readings. In the case of the gasoline prices, the reading 66.5 appears the most times (3), and is accordingly the mode. In this situation, a case could be made for promoting 66.5¢/*l* as the most representative value.

Median

The *median* is the central value in the set of readings. To find the median, first set down the readings in descending order of magnitude. Then cross off the highest and lowest readings, then the next highest and next lowest, and so on, until there is only one reading left. That reading will be the median. Its significance is that there will be as many readings that are greater than the median, as there are less than the median.

For the set of 15 gasoline prices, the median turns out to be 65.5¢/*l*.

One data assembling situation in which the median is often used is in the recording of salaries. If a person is earning the median salary, then half of the people reporting are earning less than he or she is, and the other half are earning more. The compilers of the data will sometimes list the salaries reported in descending order of magnitude, and then divide the list into 10 equal groups. When this is done the groups are called *deciles*. The desirable situation, of course, is for one to be in the upper decile. If the same list is divided into 4 groups instead of 10, then the groups are called *quartiles*.

In general, there are no hard and fast rules that decide which of the four possibilities should be chosen as the most representative value. The particular situation itself governs.

Predicting Future Values

Sometimes the purpose of determining the most representative value is to estimate what that value will be at some future time. Preparing a budget is a prime example of this task. For example, suppose that a car owner has the following data for what he has spent on gasoline over the last five years.

Table 11-1. History Data for Predicting Future Values

Years Ago	$ Spent for Gasoline
5	850
4	680
3	810
2	1070
1	1060

Based on these numbers, how much should be budgeted for gasoline this year?

From the appearance of the numbers, the best and easiest way out of the dilemma would be to determine the weighted average, which turns out to be $950. However, to get some idea of any trend that may be developing, it is necessary to plot the values on a time base, as in Figure 11-1. In this graph, the negative signs indicate years in the past.

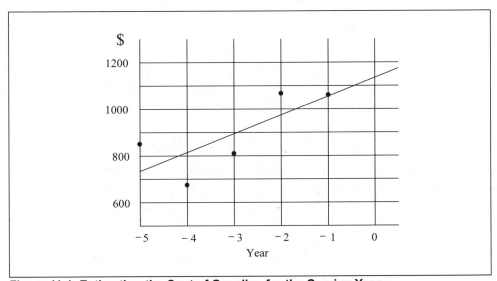

Figure 11-1. Estimating the Cost of Gasoline for the Coming Year.

Unfortunately, the way the points are dispersed makes it difficult to eye-ball in the best straight line, but help is on the way by virtue of a mathematical procedure that involves the use of calculus, but which we need not go into here. The equation of the best line will have the form y = a + bx. The procedure assumes that as far as the various points are concerned, the x values are correct (no error), but the y values deviate from their logical values represented by the best line. Expressions can then be devised for the constants a and b based on minimizing the sum of the squares of all of the y deviations.

These expressions turn out to be:

$$a = \frac{\sum x \sum xy - \sum y \sum x^2}{\left(\sum x\right)^2 - n \sum x^2}, \text{ and } b = \frac{\sum x \sum y - n \sum xy}{\left(\sum x\right)^2 - n \sum x^2}$$

where n is the number of points, and Σ is the summation sign.

This theory should be applied to the data on gasoline costs. From the five data points obtained:

Table 11-2. Predicting Future Values Example

	x	y	xy	x^2
	−5	850	−4 250	25
	−4	680	−2 720	16
	−3	810	−2 430	9
	−2	1 070	−2 140	4
	−1	1 060	−1 060	1
Σ	−15	4 470	−12 600	55

$$\therefore a = \frac{[-15 \times -12\ 600] - [4\ 470 \times 55]}{[-15^2] - [5 \times 55]} = 1\ 137$$

$$b = \frac{[-15 \times 4\ 470] - [5 \times -12\ 600]}{[-15^2] - [5 \times 55]} = 81$$

The equation for the best line is consequently y = 1 137 + 81 x.

The purpose of obtaining the equation for the best line was to estimate the value of y, the cost of gasoline, at x = 0, which corresponds to the current year. When x = 0, y is equal to $1 137, although judging from the original data from which the line was determined, four significant figures are hardly justified.

In summary, using the data on the cost of gasoline over the previous five years was useful for demonstrating the procedure for establishing the best straight line through a collection of points. However, as a means of estimating how much gasoline is going to cost in the current year, it is obviously a less reliable method than say, determining the weighted average. The best straight line method using least squares is better applied to collections of points that the laws of nature say *should* lie on a straight line, but do not, due to inaccuracies in the data.

If the best line procedure is used, one should take care to distinguish between Σx^2 and $(\Sigma x)^2$.

How Much Confidence in the Most Representative Value?

It often happens that the most representative value (MRV) for a particular entity has to be determined from a number of measurement readings, with the added complication that not all of the readings are the same. Differences in the readings can occur for a number of reasons.

- The entity varies from time to time, as in the case of gasoline prices, or the outdoor temperature, but one reading has to be selected for the MRV to make cost or other projections.

- Not all of the readings are taken by the same person, and there is a question of skill involved.

- Not all of the readings are taken using the same type or quality of equipment, and there is a question of the potential error or the reliability.

The result is that when one number is designated to be the MRV out of a group of readings representing the same entity, the question then arises: How much assurance can one have in the accuracy of the MRV?

A possible answer lies in determining how closely the various readings are grouped around the MRV, or conversely, how badly they are scattered. In this procedure, the MRV is, by definition, the arithmetic mean (AM) or average. For the set of gasoline prices (P1 to P15), the AM is 65.2¢/l. The deviations of the individual readings are designated D1, D2,…D15, where

$$D1 = P1 - AM, D2 = P2 - AM… D15 = P15 - AM.$$

All deviations (D1 to D15) are considered to be positive even if the AM is greater than an individual price reading. The next step is to calculate the *average deviation*, which is the sum of all the deviations divided by the number of readings. For the gasoline price example, the average deviation

$$AD = \frac{2.7+0.7+0.3+1.3+3.7+3.7+2.7+1.3+2.3+2.3+3.3+3.3+1.3+0.3+1.7}{15}$$

$$= 2.1.$$

The average deviation, in itself, is a fairly good indicator of the confidence one can place in the arithmetic mean as the MRV. If the price of gasoline is stated to be $65.2 \pm 2.1\text{¢}/l$, it implies that the price posted on any particular day has a 50% chance of being in the bracket 63.1 to 67.3¢/l.

The Standard Deviation

The standard deviation is a number developed by statisticians to indicate the extent of the dispersion of the readings around the arithmetic mean. The standard deviation is not the same as the average deviation, although the average deviation is involved in the calculation of the standard deviation.

Assuming that the average deviation AD has been determined as described already, then the standard deviation SD is equal to

$$\sqrt{\frac{\Sigma(D-AD)^2}{n-1}}.$$

This is the abbreviated version, in which Σ is the summation operator, D stands for the deviations of the individual readings from the AM, AD is the average deviation, and n is the number of readings. In the long form,

$$SD = \sqrt{\frac{(D1-AD)^2 + (D2-AD)^2 + (D3-AD)^2 + \dots + (D15-AD)^2}{n-1}}.$$

For the gasoline price situation, the standard deviation computes to be 1.7.

Statistically, all of the readings that were involved in the calculation should be not more than three times the standard deviation different from the arithmetic mean. If any individual reading has a deviation greater than this from the AM, it is assumed to be invalid and it is scrubbed from the list. The calculation is then redone, from the start, using the remaining readings. This will generate a new value for the standard deviation and will require a second check of the validity of the readings. For the example, all 15 readings were within $3 \times 1.7 = 5.1\text{¢}/l$ of the AM, 65.2¢/l, and are accordingly accepted as valid.

The statistical significance of the standard deviation is that of all of the readings that were taken, or which will subsequently be taken, 68% of the readings will lie within the bracket AM plus or minus the standard deviation, 95% will be in the bracket AM plus or minus twice the standard deviation, and 99% will be in the bracket AM plus or minus three times the standard deviation.

Stated in another way, if the standard deviation can be determined from a set of readings, then it can be expected that 99% of the time, future readings will be within three times the standard deviation of the AM, 95% of the time they will be within twice the standard deviation of the AM, and 68% of the time they will be within the bracket AM plus or minus the standard deviation. This implies a certain level of assurance.

When the MRV for an entity is stated, it is usually important to know the confidence level that should be given to the statement. If, for example, the price of gasoline is said to be 65.2¢/l at the 99% confidence level, it will mean that the potential error should be considered to be plus or minus three times the standard deviation.

The validity of this statistical theory depends heavily on the number of readings on which the calculations are based. Generally, the more, the better, although a point of diminishing returns can be reached. Fewer than 10 readings in most cases are inadequate.

Curve Fitting

It sometimes happens that a computer needs to know the relationship between two variables, but the relationship unfortunately does not follow any particular mathematical law. That is, the relationship is not parabolic, hyperbolic, sinusoidal, exponential, logarithmic, or whatever. Physical properties of naturally occurring substances often fall into this category. Enough data are available that the dependent variable can be plotted against the independent variable, but unfortunately, computers are not very good at reading graphs.

The graph in Figure 11-2 shows the variation of the vapor pressure of water with temperature. The problem is, given any temperature between 0 and 100°C, to have a computer come up with the correct value of the vapor pressure.

This can be done through the use of a curve fitting program. The general equation for the curve fit has the form

$$y = f(z) = I_0 + I_1 z + I_2 z^2 + I_3 z^3 + I_4 z^4$$

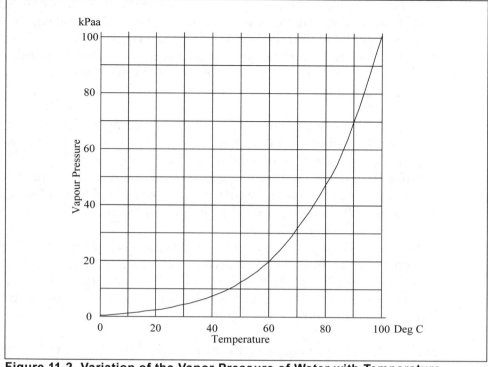

Figure 11-2. Variation of the Vapor Pressure of Water with Temperature.

where the I's are appropriate constants to be determined, and z is an intermediate variable, which will be clarified shortly.

The procedure is as follows.

1.   Divide the range of the independent variable (x) into four equal sectors. The division points are identified as x_0 (the starting point), $x_1, x_2, x_3,$ and x_4 (the end point). If each sector is w units wide, in units of the independent variable, then

$$w = \frac{x_4 - x_0}{4}.$$

2.   Record the values of the independent variable (y), which correspond to $x_0, x_1, x_2, x_3,$ and x_4. Call these $y_0, y_1, y_2, y_3,$ and y_4 The required values of the constants $I_0, I_1, I_2, I_3,$ and I_4 will then be:

$$I_4 = \frac{y_4 - 4y_3 + 6y_2 - 4y_1 + y_0}{24}$$

$$I_3 = \frac{y_3 - 3y_2 + 3y_1 - y_0}{6} - 6I_4$$

$$I_2 = \frac{6y_2 - 9y_1 + 4y_0 - y_3}{6} - 2I_3 - I_4$$

$$I_1 = y_1 - y_0 - I_2 - I_3 - I_4$$

$$I_0 = y_0.$$

3.   The intermediate variable

$$z = \frac{x - x_0}{w}$$

where x is the value of the independent variable at which the value of the dependent variable y is to be calculated. Then,

$$y = I_0 + I_1 z + I_2 z^2 + I_3 z^3 + I_4 z^4.$$

Table 11-3 shows how well the curve fit values compare with the actual vapor pressure values. The curve fit graph has not been plotted in Figure 11-2, since except for a small region around 10°C, it coincides almost exactly with the real vapor pressure curve.

Note that the curve fit values are dead accurate at the 0, 25, 50, 75, and 100% points. This is an inherent characteristic of the curve fit program and will apply any time it is used. If greater accuracy is required, the graph can be divided into two sections and a curve fit equation developed for each section. The combination will then be exactly accurate at nine points along the graph instead of five.

The curve fit program can be applied to graphs that change direction, such as process reaction curves, which have the characteristic S shape. The accuracy of the fit in these cases is usually quite good.

An important word of caution: The curve fit equation should never be extrapolated outside of the specified lower and upper limits (x_0 and x_4). The accuracy falls off rapidly outside of these limits.

Table 11-3. The Curve Fit Program

Temperature (x) Deg C	Vapor Pressure kPa	Curve Fit Value (y)
0	0.60	0.60
10	1.22	1.13
20	2.33	2.30
25	3.16	3.16
30	4.23	4.26
40	7.35	7.38
50	12.29	12.29
60	19.85	19.82
70	31.06	31.04
75	38.43	38.43
80	47.20	47.25
90	69.89	70.00
100	101.03	101.03

Index